ECONOMIC AND SOCIAL COMMISSION
FOR ASIA AND THE PACIFIC
Bangkok, Thailand

ENERGY ISSUES AND PROSPECTS IN THE ASIA AND PACIFIC REGION

ENERGY RESOURCES DEVELOPMENT SERIES
No. 31

UNITED NATIONS
NEW YORK, 1988

ST/ESCAP/623

UNITED NATIONS PUBLICATION

Sales No. : E. 88.II.F.9

ISBN 92-1-119463-6

ISSN 0252-4368

01550

FOREWORD

This is the third publication in the Series of Energy Resources Development Problems in the ESCAP region. The first one dealt with general energy issues covering all sources of energy, whereas the second dealt exclusively with new and renewable energy issues. This publication contains the report of, and information and issue papers on energy presented to the Committee on Natural Resources at its fourteenth session held at Bangkok in 1987.

The Committee on Natural Resources, a legislative body of the Economic and Social Commission for Asia and the Pacific, meets on a biennial basis to study, review and evaluate progress in natural resources development in the region, in particular, in the fields of: energy; water; mineral; marine resources and cartography and remote sensing. The Committee, at its fourteenth session, dealt with various energy issues in great detail.

The three main themes that were among the subjects of the Committee's deliberations were: the current energy situation and demand management achievements; energy issues comprising prospects for production and utilization of coal, natural gas and electricity, and human resources development; and strategies required in the accelerated development of new and renewable sources of energy. As usual a number of information and issue papers were presented under each theme of the Committee. Staff members of the ESCAP secretariat prepared most of the issue papers. A number of country delegations and other international organizations also presented papers.

The first chapter of the publication contains the report of the Committee on its extensive deliberations in the fields of energy and natural resources. A summary of conclusions and recommendations of the Committee, geared to accelerate the development and supply of natural resources for economic growth, has also been included in this chapter.

Various key papers presented to the Committee highlighting salient energy issues have been included in chapter II of the publication. The papers have been grouped according to the main themes mentioned above. Wherever possible full texts with minor editing have been retained. In other cases, owing to limitation of space, either a synopsis or an abridged form of original reports/papers have been incorporated. The secretariat wishes to acknowledge the contributions of all concerned; particular mention could be made of the two theme papers contributed by the International Labour Organisation and the Energy Research Group of the International Development Research Centre.

CONTENTS

ANNEX

I. REPORT OF THE COMMITTEE ON NATURAL RESOURCES ON ITS FOURTEENTH SESSION

A. Summary of conclusions and recommendations*

The significant conclusions and recommendations of the Committee are highlighted below.

The Committee:

(1) Expressed its appreciation to the secretariat for having prepared an excellent document not only covering the current energy scene but also providing an analysis of expected trends in commercial energy demand, and recommended that the secretariat should continue to prepare such reports in the future.

(2) Expressed its concern over the future availability of oil resources and recommended that, to make existing resources last longer, all countries had to adopt measures for energy conservation and energy demand management by efficient utilization of energy and through reducing energy intensity. However, it felt that, since the per capita energy consumption was much lower in developing than in industrialized countries, the gap should be narrowed through increased energy production, while at the same time avoiding unnecessary waste.

(3) Recognized that new and renewable sources of energy had been playing an important role in rural areas. However, it voiced concern over the alarming extent of deforestation, soil degradation and erosion resulting from the increased use of fuelwood. It recommended, therefore, that the use of alternative sources wherever available, such as small hydropower, coal from small-scale mining, gas, solar energy and biogas, should be enhanced.

(4) Requested that ESCAP continue to provide advisory services and technical assistance in various fields of energy.

(5) Generally endorsed the conclusions and recommendations of various studies mentioned in the reports (on regional energy economy patterns and the effects of price and non-price policies on energy demand management).

(6) Observed that current low energy prices hindered development conservation of energy, but recognized the importance of energy conservation. It felt that if stability was achieved in the energy markets, it would bring benefits to all.

(7) Stressed the necessity of strengthening international co-operation in building up statistics on sectoral energy demand.

(8) Endorsed the views expressed in the documentation on the need for intensive development of energy supply systems of coal, natural gas and electricity to cope with the forecast rapid increase in energy demand in view of the limited supply of oil.

(9) Noted that interconnection between power systems was an effective means of improving the overall level of system stability and economy.

(10) Recognized the important and critical role of manpower planning in the promotion of the economic and social development of developing countries in the ESCAP region and endorsed the TCDC working group concept proposed by the secretariat.

(11) Generally endorsed the secretariat's approach to co-operative research, development and demonstration in the field of new and renewable sources of energy, namely, integrating those activities into overall national energy development activities and promoting co-operation through a tripartite review conference and a networking mechanism.

(12) Appreciated that extensive activities had been carried out by the secretariat in the energy sector and hoped that it would continue such work for the following years.

(13) Considered and endorsed four reports – (E/ESCAP/NR.14/12), containing a description of activities undertaken in the water resources sector since the thirteenth session of the Committee, E/ESCAP/NR.14/13, on rainwater harvesting techniques and prospects for their application in developing island countries, E/ESCAP/NR.14/16, on water tariffs as a policy instrument to provide better management of water resources, and E/ESCAP/NR.14/21, on safety evaluation of dams.

(14) Commended the progress achieved by the secretariat in the implementation of the work programme in the water resources sector.

(15) Commended the substantial progress achieved by the secretariat the implementation of the work programme in the mineral sector, including the training activities in various areas of mineral resource exploration and development and promotion of technical co-operation.

(16) Commended the secretariat on its role in organizing an entirely new area of ESCAP assistance in urban geology and stressed the need to continue the programme of pilot studies of urban geology of the large population centres of the region.

(17) Appreciated the efforts and endorsed the activities of the secretariat under the regional mineral resources

---

* This summary does not form part of the report adopted by the Committee.

development programme. To enable those activities to be carried out, it urged UNDP to seriously consider the funding of the project on the programme.

(18) Supported the secretariat's effort to integrate activities in the field of marine resources in keeping with the spirit of the United Nations Convention on the Law of the Sea and related resolutions, and endorsed its activities in that area, especially with respect to the provision of relevant information.

(19) Endorsed all activities of the regional remote sensing programme and strongly supported its co-ordinating role in the organization of training, seminars, workshops and meetings; the sponsoring of pilot studies and technical co-operation exchanges; and the regional information system (RIS).

(20) Recommended that, in addition to the use of regional training facilities, such as those at the Asian Institute of Technology, the national facilities for training should be utilized more.

(21) Recognized the need to start an ESCAP data bank on remote sensing to be included in the regional information system and recommended that information be disseminated as a part of the *Remote Sensing Newsletter.*

(22) Appreciated the presentation of the draft medium-term plan, 1990-1995, for the energy programme and agreed with the approach for meeting the felt regional needs. It urged the secretariat to undertake the necessary steps, including mobilizing adequate finance, to implement the plan.

(23) Endorsed the proposed draft medium-term plan, 1990-1995, in the field of water resources.

(24) Endorsed the secretariat's initiative in presenting the subject of flood loss prevention and mitigation as a major issue to be presented to the Commission for consideration at its forty-fourth session.

(25) Suggested that the Commission be requested for further guidance on the subject of shared water resources at its forty-fourth session, since there was neither conclusion nor consensus on the subject at the current session of the Committee.

(26) In considering the draft medium-term plan, 1990-1995, on mineral resources, recommended that the study of the region's sedimentary basins focus on their natural resources, especially on hydrocarbon, ground water, coal and non-metallic minerals, and that those studies be co-ordinated with other ongoing sedimentary basins resource assessment programmes.

(27) Expressed its support of the draft medium-term plan, 1990-1995, on marine affairs.

(28) Requested that a report on the use of cartography and remote sensing for resources evaluation and environmental protection be included in the draft provisional agenda for the fifteenth session.

(29) Endorsed the draft provisional agenda (with some additions) for the fifteenth sension of the Committee.

B. Organization of the session

The Committee on Natural Resources held its fourteenth session at Bangkok from 27 October to 2 November 1987.

Attendance

The session was attended by representatives of the following ESCAP member countries: Australia, Bangladesh, China, Democratic Kampuchea, France, India, Indonesia, Islamic Republic of Iran, Japan, Malaysia, Netherlands, New Zealand, Pakistan, Philippines, Republic of Korea, Sri Lanka, Thailand, Union of Soviet Socialist Republics, United Kingdom of Great Britain and Northern Ireland, United States of America, and Viet Nam.

Representatives of Finland, the Federal Republic of Germany, and Norway also attended the session, in accordance with paragraph 9 of the terms of reference of the Commission.

The Departments of International Economic and Social Affairs and of Technical Co-operation for Development of United Nations Headquarters, and the following United Nations bodies and specialized agencies were represented: United Nations Development Programme (UNDP), United Nations Environment Programme (UNEP), International Labour Organisation (ILO), Food and Agriculture Organization of the United Nations (FAO), World Health Organization (WHO), World Bank, World Meteorological Organization (WMO) and United Nations Industrial Development Organization (UNIDO). Observers from the following also attended: Asian Institute of Technology (AIT), International Development Research Centre (IDRC), International Energy Agency/Organisation for Economic Co-operation and Development (IEA/OECD), International Union of Geological Sciences (IUGS), South Pacific Bureau for Economic Co-operation (SPEC), and World Energy Conference (WEC).

Opening of the session

The Deputy Executive Secretary of ESCAP delivered the statement of the Executive Secretary and opened the session.

In his statement, the Executive Secretary attached greater importance than usual to the Committee, since at the current session it would be deliberating on energy issues that had originally been planned for discussion at a ministerial meeting, and would be instrumental in mapping the work of the secretariat in the sphere of energy up to 1995.

He stressed that despite the current relatively low price of oil, there was no reason to think that the energy problem was over. As the known oil resources were finite, he asked all concerned to work together to find alternative sources of energy to meet the ever-increasing demand. In the field of commercial energy supplies, coal, natural gas and electricity were expected to play an increasingly greater role because of oil resource constraints. Such issues as institutional aspects, trade, and the building up of adequate infrastructure were worth considering in order to facilitate and enhance the alternative energy substitution process. Human resources development for the increasingly complex sector of energy was another critical issue.

Turning to the issue of rural energy, the Executive Secretary pointed out that traditional energy still contributed over 50 per cent of energy supplies in the developing countries of the region and was crucial to and often the only available source for rural dwellers. To meet expanded rural needs for energy, he stressed that accelerated development of new and renewable sources of energy was very important. Some of the common problems in the energy management of developing countries could be solved more effectively through co-operative research, development and demonstration among countries of the region.

In highlighting the secretariat activities in the field of energy, the Executive Secretary mentioned that ESCAP activities continued to be reinforced by the three regional programmes: the regional energy development programme (REDP), the Pacific energy development programme (PEDP) and the regional network on biomass, solar and wind energy (BSW). He reiterated ESCAP appreciation to UNDP for generously funding REDP and PEDP, and to the Governments of Japan and Australia for supporting the BSW project.

The Executive Secretary also referred to the excellent co-operation the ESCAP secretariat enjoyed with other international organizations, agencies, and institutions involved in energy activities in the region.

He then briefly highlighted some issues and concerns about other sectors of natural resources: water, mineral and marine resources, and cartography and remote sensing. In the water sector, pointing to the loss of lives and property caused by floods in various parts of the region, the Executive Secretary referred to the Typhoon Committee's recommendation that further work be undertaken in the field of flood protection. In the sphere of mineral resources, the Executive Secretary drew the attention of the Committee to recent developments in the mineral resources development programme.

The Executive Secretary reiterated his appreciation to donor countries and agencies that contributed generously to ESCAP activities in the natural resources and energy programmes, and hoped that such contributions would continue to flow so as to keep regional activities going.

Election of officers

The Committee elected Mr. Phol Songpongs (Thailand) as Chairman, and Mr. Nuruddin M. Kamal (Bangladesh) and Mr. Seyed Ahmad Hakim (Islamic Republic of Iran) as Vice-Chairmen. It also elected Mr. A.J. Surjadi (Indonesia) as Rapporteur and Chairman of the Drafting Committee.

Adoption of the agenda

The Committee considered and adopted the following agenda:

1. Opening of the session.

2. Election of officers.

3. Adoption of the agenda.

4. Current energy situation and demand management achievements:

   (a) Regional energy scene;

   (b) Regional energy economy patterns, including demand management, price and non-price policies and energy conservation measures.

5. Energy issues:

   (a) Prospects for production and utilization of coal, natural gas and electricity;

   (b) Human resources development.

6. Strategies required in the accelerated development of new and renewable sources of energy:

   (a) Contributions of new and renewable sources of energy to the regional energy supply;

   (b) Achievements in co-operative research, development and demonstration and future plans.

7. Activities of ESCAP in regard to natural resources:

   (a) Energy resources (other than those activities referred to above under energy and other programmes);

   (b) Water resources;

   (c) Mineral resources;

   (d) Marine resources;

   (e) Cartography and remote sensing.

8. Activities of other international bodies in the appraisal, development and management of energy resources.

9. Programme changes, 1988-1989, and proposed draft medium-term plan, 1990-1995;

    (a) Energy resources;

    (b) Water resources;

    (c) Mineral Resources;

    (d) Marine resources;

    (e) Cartography and remote sensing.

10. Consideration of the agenda and arrangements for subsequent sessions of the Committee.

11. Other matters.

12. Adoption of the report.

C. Current energy situation and demand management achievements
(Item 4 of the agenda)

1. Regional energy scene

The Committee had before it document E/ESCAP/NR.14/9, prepared by the secretariat, highlighting the current energy situation in the Asian and Pacific region in respect of resources, production, consumption and demand. The document contained an analysis of the resource situation and current scene in different forms of energy, such as solid fuels, liquid fuels, gaseous fuels and electricity, as well as non-commercial forms of energy. Concerning the regional scene in commercial energy demand, it contained an analysis of recent trends in energy consumption and consumption patterns followed by a projection of future trends and patterns in commercial energy demand in selected countries of the ESCAP region.

The Committee noted with interest the information provided by various country representatives, through their country papers and statements, on the latest energy situation, trends in development and management of energy resources as well as policies and issues in the energy sector of their respective countries.

The Committee expressed its appreciation to the secretariat for having prepared an excellent document not only covering the current energy scene but also providing an analysis of expected trends in commercial energy demand. It recommended that the secretariat should continue to prepare such reports in the future. Some updated figures were provided by a few countries.

The Committee recognized the high potential for hydropower which remained unexploited in the region, as highlighted in the secretariat paper. It appreciated the inclusion of information on small, mini and micro hydropower resources for the first time in the secretariat's review of regional hydropower development.

The Committee expressed concern over the future availability of oil resources and recommended that, to make existing resources last longer, all countries had to adopt measures for energy conservation and energy demand management by efficient utilization of energy and through reducing energy intensity. However, it felt that, since the *per capita* energy consumption was much lower in developing than in industrialized countries, the gap should be narrowed through increased energy production, while at the same time avoiding unnecessary waste.

The Committee noted the efforts made by the countries to diversify the use of energy resources in the substitution process from a largely oil-based supply to a mix of alternative sources of energy, such as coal, natural gas, hydropower and, to some extent, nuclear energy.

The Committee also noted the high growth of demand for electric power in the developing countries of the region and the efforts that various countries had been making to meet it.

The Committee recognized that new and renewable sources of energy had been playing an important role in rural areas. However, it voiced concern over the alarming extent of deforestation, soil degradation and erosion resulting from the increased use of fuelwood. It recommended, therefore, that the use of alternative sources wherever available, such as small hydropower, coal from small-scale mining, gas, solar energy and biogas, should be enhanced.

The Committee requested that ESCAP continue to provide advisory services and technical assistance in various fields of energy.

The Committee was informed of the bilateral and multilateral assistance provided by France, Japan and the USSR to some developing countries and regional activities.

2. Regional energy economy patterns, including demand management, price and non-price policies and energy conservation measures

The Committee had before it documents E/ESCAP/NR.14/1, entitled "Issues related to the regional economy pattern", and E/ESCAP/NR.14/2, entitled "Effects of price and non-price policies on energy demand management".

While the first document discussed the results and recommendations of policy analysis studies presented at the Asian Forum on Energy Policy (27-30 October 1986), the second presented a three-year implementation plan for giving assistance to the strengthening of national capabilities for energy planning and data management under the ESCAP regional energy development programme (REDP).

After complimenting the secretariat on the quality of the documentation, the Committee generally endorsed the conclusions and recommendations of various studies mentioned in the reports. Some countries expressed interest in participating in REDP activity P-1.1, "Sectoral energy demand studies".

The Committee noted with appreciation the support to activities in sectoral demand analysis by UNDP, by the Federal Republic of Germany, through the German Agency for Technical Co-operation (GTZ), and also by the Government of France for providing the lead consultant for REDP activity P-1.1.

The Committee observed that current low energy prices hindered development and conservation of energy, but recognized the importance of energy conservation. It felt that if stability was achieved in the energy markets, it would bring benefits to all.

The Committee stressed the necessity of strengthening international co-operation in building up statistics on sectoral energy demand.

The Committee took note of energy conservation efforts, including policy measures, undertaken in a number of countries in the region. The Committee was informed that such measures were effective in many countries.

D. Energy issues
(Item 5 of the agenda)

1. Prospects for production and utilization of coal, natural gas and electricity

The Committee had before it documents E/ESCAP/NR.14/3, entitled "The use of coal in households and small-scale industries", E/ESCAP/NR.14/4, entitled "Report of the Meeting of Senior Experts Preparatory to the Fourteenth Session of the Committee on Natural Resources", E/ESCAP/NR.14/7 and Corr.1, entitled "Potential for Asian trans-country power exchange and development", and E/ESCAP/NR.14/10, entitled "Prospects for production and utilization of coal, natural gas and electricity".

With regard to the identification of energy issues, the key document was the report of the Meeting of Senior Experts Preparatory to the Fourteenth Session of the Committee on Natural Resources, which was relevant to the rest of the agenda items (including item 5 (b) on human resources development), leading to a logical formulation of the medium-term plan in energy under item 9 of the agenda, after having discussed all the issues. Policy studies completed for the analysis of issues were presented in those papers concerning coal, natural gas and electricity, the use of coal for meeting dispersed domestic and small industrial needs being highlighted separately in E/ESCAP/NR.14/3, while the economies of scale obtainable through linking national electric grids were introduced in E/ESCAP/NR.14/7 and Corr.1.

The Committee endorsed the views expressed in the documentation on the need for intensive development of energy supply systems of coal, natural gas and electricity

to cope with the forecast rapid increase in energy demand in view of the limited supply of oil.

Many country representatives described their national coal, natural gas and electricity development programmes, aimed at utilizing indigenous resources as much as possible. In that connection, India expressed willingness to share its experience in the development of labour-intensive technology for medium-sized coal mines developed by that country. Australia also offered assistance in sharing its experience and technology in coal development and utilization, while emphasizing that lowest cost deposits, wherever those could be found, should be developed first.

The representative of New Zealand also described its energy activities of interest to the Committee, specifically the experience of its synthetic gasoline project and programmes on the use of CNG for transport fuel, as well as current efforts towards "decontrol" of energy markets to promote overall efficiency, new and renewable energy achievements (in biogas and geothermal energy), and successful energy conservation policies.

Based on the assumption of the need for a tremendous amount of coal to be burnt and anticipating a resulting increase in pollution problems, the suggestion was made that the secretariat should organize a meeting to exchange experience on that issue. It was also suggested that a meeting on maximizing the use of natural gas as chemical feedstock, for cooking etc., should be organized, taking into account the investment requirement for development, advantages in pollution control, and so on. International co-operation in research on that issue was also suggested.

The Committee noted that interconnection between power systems was an effective means of improving the overall level of system stability and economy, and that the ASEAN Electricity Network Interconnection Group had been active in promoting an ASEAN grid. Some corrections were made in the secretariat paper based on updated factual information.

One delegation expressed the view that his country preferred to have such projects undertaken on a bilateral basis rather than under the auspices of any multilateral agency.

2. Human resources development

The Committee had before it document E/ESCAP/NR.14/8 prepared by the secretariat and a conference room paper prepared by ILO, dealing with human resources development issues in the energy sector. The secretariat paper addressed the manpower planning issue through an objective assessment and evaluation of the areas where energy training activities were urgently needed and the types of training required for further human resources development. The paper elaborated on the TCDC (technical co-operation among developing countries) working group concept and the role of regional and subregional co-operation in energy manpower training.

The ILO paper dealt with the difficulties of manpower planning in a climate of uncertainty and structural change. It pointed out that that human dimension of energy planning was an important issue and that the manpower issue today should include a search for flexibility and a capacity to adjust to changes and external shocks, and to solve problems so as to minimize the risks to expensive capital investment and installations.

The Committee commended the secretariat on the preparation of the paper and recognized the important and critical role of manpower planning in the promotion of the economic and social development of developing countries in the ESCAP region. Many representatives commended the secretariat's work on manpower training in the energy sector, and expressed continued support for its efforts towards upgrading manpower capabilities. The Government of Japan informed the Committee that it would host an expert group meeting on integrated action plans in Tokyo from 7 to 10 December 1987. The TCDC working group concept proposed by the secretariat was endorsed by the Committee and some delegations expressed willingness to participate in such working group arrangements.

E. Strategies required in the accelerated development of new and renewable sources of energy
(Item 6 of the agenda)

1. Contributions of new and renewable sources of energy to the regional energy supply

The Committee had before it document E/ESCAP/NR.14/5, containing an assessment of the contribution of new and renewable sources of energy to regional energy supply.

The paper emphasized that satisfying regional energy demand would require a mix of renewable and non-renewable resources. The Committee recognized the importance of the contribution of new and renewable sources of energy to energy supply, particularly to the rural areas. Many representatives emphasized the need to accelerate the utilization and development of new and renewable sources of energy. Small hydropower, biomass, wind energy and solar photovoltaic technology were considered economically and technically feasible at present. The Committee endorsed the recommendations put forward by the secretariat.

The Committee considered that the secretariat's paper gave a comprehensive overview of the scenario of new and renewable sources of energy in the ESCAP region.

Many delegations reported the progress of their new and renewable energy development programmes and

distributed their country reports. Economic applications as well as socio-economic benefits were emphasized, while further cost reductions of some already commercially mature technologies were expected through continued deployment efforts.

2. Achievements in co-operative research, development and demonstration and future plans

The Committee had before it document E/ESCAP/NR.14/6, entitled "Co-operative research, development and demonstration achievements and future plans regarding new and renewable sources of energy", and also a publication entitled *Energy research – Directions and Issues for Developing Countries* by the Energy Research Group convened and supported by the International Development Research Centre (IDRC) and the United Nations University (UNU).

Document E/ESCAP/NR.14/6, while giving a general overview of the role of new and renewable sources of energy programmes and major issues for the medium-term plan period 1990-1995 and beyond, also reviewed ESCAP achievements in promoting co-operative research, development and demonstration through the regional network on biomass, solar and wind energy. Concrete co-operative programmes were presented in the paper as a recommendation for continued tripartite co-operation among developed countries, developing countries, and regional institutes.

The Energy Research Group document stressed the multi-sectoral and all-pervasive nature of energy problems, listing six areas for the integration of energy research with economic development issues that were of particular relevance to Asian countries. Those areas included both conventional and renewable energy research directions.

The Committee expressed appreciation of both reports, and generally endorsed the secretariat's approach to co-operative research, development and demonstration in the field of new and renewable sources of energy, namely, integrating the activities concerned into overall national energy development activities and promoting co-operation through a tripartite review conference and a networking mechanism.

It was suggested that more emphasis should be placed on ways and means of promoting exchange of equipment for testing and research purposes as well as on projects and training through exchange of research personnel among institutes in different countries. The suggestion was also made that, beyond the concept of tripartite co-operation in research, development and demonstration in new and renewable sources of energy, suitable mechanisms for promoting co-operative arrangements should be explored, and roving training courses organized.

The Committee expressed appreciation of the continued support from Australia, Japan and others of regional activities in that field, while concern about continued funding support was also expressed, and the need for sustained efforts for a viable energy transition process was emphasized.

With regard to the tripartite research, development and demonstration project on solar photovoltaic technology (funded by the Government of Japan through the Japan-ESCAP Co-operation Fund), the representative of Japan informed the Committee that his Government had noted with satisfaction that: (a) the seminar/training course held in Indonesia had been successfully completed and the training had been considered useful by many member countries; and (b) the roving training courses on photovoltaic technology for Pacific island countries, which had commenced in February 1987, were also being implemented with successful results. He also indicated his Government's concern at the delay in implementing the training activity in Pakistan. He further informed the Committee that if that project was not implemented in the near future, there would be some adverse effects on Japan's co-operation and assistance in that field. In reply to the above comment, the representative of Pakistan informed the Committee that his Government had agreed to hold the proposed training programme in the first half of 1988 and would accept an ESCAP mission to Pakistan immediately to finalize arrangements for the programme. He also conveyed his Government's request that the requisite photovoltaic equipment be delivered directly to Islamabad.

F. Activities of ESCAP in regard to natural resources
(item 7 of the agenda)

1. Energy resources

The Committee had before it document E/ESCAP/NR.14/18, which contained a brief review of the activities of the secretariat on energy resources carried out since the report to the Committee on its thirteenth session in 1986. During that period, upon the request of member countries, extensive advisory services had been provided in both conventional and new and renewable sources of energy and a number of studies had been produced. In total, 21 missions, 9 meetings, seminars and workshops, and 3 training courses had been carried out and 22 reports and publications issued.

The Committee appreciated that extensive activities had been carried out by the secretariat and hoped that it would continue such work for the following years.

The Committee noted with appreciation the financial support of UNDP and the Governments of Australia, France, Japan, the Netherlands, and the provision of host facilities by Bangladesh, India, Indonesia, Malaysia and the Republic of Korea, as well as other contributions from various countries and organizations.

The Committee was informed that subsequent to the preparation of the above report, the PEDP document was approved on 7 August 1987 and the REDP Tripartite Review Conference was held from 26 to 28 August 1987.

Recognizing the comprehensive energy programme that ESCAP provided, many delegations asked that either REDP continue beyond its second cycle (1987-1991) or similar programmes be conducted at the end of the cycle.

2. Water resources

In the field of water resources, the Committee had before it document E/ESCAP/NR.14/12, containing a description of activities in water resources undertaken since the thirteenth session of the Committee; E/ESCAP/NR.14/13, containing a report on rain-water harvesting techniques and prospects for their application in developing island countries; E/ESCAP/NR.14/16, reporting on water tariffs as a policy instrument to provide better management of water resources; and E/ESCAP/NR.14/21, presenting methodology recommended for use in safety evaluation of existing dams. The Committee commended the work undertaken by the Section.

The activities in that sphere during the year included 5 missions, 9 meetings, seminars and workshops, and the issuance of 11 reports and publications under four programme elements.

The report on water tariffs as a policy instrument to provide better management of water resources was reviewed by the Committee. Two countries provided up-to-date information on the water charges and capital costs of irrigation and water supply projects in their respective countries.

The Committee considered the report on safety evaluation of existing dams. One delegation reported on the recent developments in that area in its country and advised that a standing committee to review practices and make recommendations for uniform dam safety practices had submitted its reports. Another country pointed out the importance of seismic risk in dam safety. One delegation indicated its country's need for assistance in monitoring the safety of existing dams.

The report on rain-water harvesting techniques and prospects for their application in developing countries was commended by the Committee as it contained information on the potential and use of rain-water harvesting for domestic water supply in island countries and on the design considerations for roof rain-water harvesting.

Two delegations presented an account of the progress achieved in the development of water resources in their respective countries. One country provided the secretariat with a list of references on ground-water pollution related to human settlements. The representative of Japan indicated that the information presented in the reports would be useful for future considerations in its planning in the water sector, and confirmed its continuing support to the secretariat, in particular on activities related to the Typhoon Committee. The USSR, reporting on its active co-operation with ESCAP in the water sector, informed the Committee of its willingness to co-operate further with the secretariat, in particular on provision of training in that sector and assistance in the formulation of water master plans.

The Committee endorsed the reports in the four documents.

3. Mineral resources

The Committee had before it documents E/ESCAP/NR.14/14 and E/ESCAP/NR.14/20, containing a summary of the activities of the secretariat in the implementation of the current work programme in the mineral sector since the thirteenth session of the Committee, and recent and future activities of the secretariat under the regional mineral resources development programme since the transfer to the ESCAP secretariat of the functions involved as from 1 March 1987.

The Committee commended the substantial progress achieved by the secretariat in the implementation of the work programme in the mineral sector, including the training activities in various areas of mineral resource exploration and development and promotion of technical co-operation. It expressed satisfaction with the activities undertaken in the appraisal of geology and distribution of mineral and hydrocarbon resources in the region, the compilation of geological and thematic maps, and the preparation of technical publications on special topics.

The Committee commended the secretariat on its role in organizing an entirely new area of ESCAP assistance, "geology for urban development", and thanked the Governments of China and the Netherlands for funding the recently-completed highly successful Expert Working Group Meeting-cum-Workshop on the Urban Geology of Coastal Areas, held at Shanghai, China in August 1987. The Committee expressed the need to continue the programme of pilot studies of the urban geology of the large population centres in the region, noting that their geologic hazards and foundations should be understood in advance of development and in order to make human settlements compatible with their natural environment. In that connection, the representative of China informed the Committee that that country would contribute a geologic study and maps of Ningbo City and a study of land subsidence in Shanghai to the ESCAP Atlas of Urban Geology series. Finally, the Committee recommended that the secretariat take the steps necessary to seek substantive support for that new activity by seeking donors and utilizing expertise already existing within and outside the region.

The Committee appreciated the efforts and endorsed the activities of the secretariat under the regional mineral resources development programme. Several delegations noted with appreciation a number of technical advisory missions on different types of mineralization and various aspects of mineral resources exploration and development undertaken by the secretariat under the programme.

The Committee agreed with the development objectives and considered that the scope of activities under the programme should be limited to the indicated number of mineral commodities in the exploration phase of mineral resources development. It requested the secretariat to seek donors urgently for experts on coal and industrial mineral exploration.

The Committee expressed its appreciation to UNDP and to China, France, Japan and the Netherlands for providing support in funding experts and training activities in the mineral sector. Moreover, it gratefully acknowledged the technical assistance provided by experts from Australia, Czechoslovakia, the Netherlands, New Zealand, the USSR and the United States, in workshops, seminars and training courses.

The Committee urged UNDP to seriously consider the funding of the project on the regional mineral resources development programme, for which the project document had been submitted to it by the ESCAP secretariat.

The Committee noted with appreciation the offer of Viet Nam to host a workshop on geophysical exploration methods for minerals in tropical rainforest areas in the near future.

The Committee also appreciated the offer of India to resume the post-graduate training course for geologists from the region and the offer of the USSR to host in 1988 a seminar on modern methods of mineral exploration. It noted with interest that the compilation of a desk study on reserves, distribution, production and potential of non-metallic minerals in the ESCAP region prepared by the USSR at the request of the ESCAP secretariat was in an advanced stage.

4. Marine resources

The Committee had before it document E/ESCAP/NR.14/17, which contained a review of the activities of the secretariat in the field of marine resources.

The Committee noted that the new regime of the sea established under the United Nations Convention on the Law of the Sea had raised national awareness of ocean-related issues in signatory member countries. The collaboration established with the Office for Ocean Affairs and the Law of the Sea in New York was commended.

The Committee supported the secretariat's efforts to integrate activities in that field in keeping with the spirit of the Convention and related resolutions.

The Committee endorsed the current activities of the secretariat in the field of marine resources. The view was expressed that the proposed study tour on marine geology and geophysical programme planning and assessment would be of benefit to participating countries. In that connection, the representative of France indicated that his Government was willing to support that activity in the near future. He also indicated that his Government would provide financial support to technical missions in South Pacific island countries, designed to assess their needs and capabilities in the field of marine affair.

As for future activities, the Committee expressed the need for secretariat assistance with respect to information on the implications of the United Nations Convention on the Law of the Sea for interested member countries and the establishment of national marine resources development and management plans.

The Committee expressed its gratitude for the continued UNDP support to the Committee for Co-ordination of Joint Prospecting for Mineral Resources in Asian Offshore Areas (CCOP) and the Committee for Co-ordination of Joint Prospecting for Mineral Resources in South Pacific Offshore Areas (CCOP/SOPAC). The Committee also noted with great appreciation the continued substantial support given by various donor countries to the programme activities of CCOP and CCOP/SOPAC, which had enabled them to assess and evaluate the mineral resources of their parts of the ESCAP region more effectively.

5. Cartography and remote sensing

The Committee had before it document E/ESCAP/NR.14/15, concerning the activities of ESCAP in cartography and remote sensing, and also including the report of the regional remote sensing programme for 1986/87. It was noted that most of those activities were being carried out under the UNDP-funded regional project RAS/86/141 (1987-1991).

The Committee endorsed all the activities of the regional remote sensing programme and strongly supported its co-ordinating role in the organization of training, seminars, workshops and meetings; the sponsoring of pilot studies and technical co-operation exchanges; and the regional information system (RIS).

The Committee was informed by one delegation, about the multi-stage and multi-sensor remote sensing techniques as constituting an efficient method for collecting data on natural resources. An integrated system for acquisition of up-to-date resource information on a multi-level remote sensing approach was recommended.

The Committee appreciated the continued support of UNDP for the project for the second phase and expressed satisfaction that UNDP had approved the project for the period 1987-1991.

The need for human resources development in the field of remote sensing was recognized by the Committee and it was recommended that in addition to the use of regional training facilities, such as those at the Asian Institute of Technology (AIT), the national facilities for training should be utilized more fully. The Committee also noted with appreciation the desire of Japan to assist the ESCAP member countries further through training projects sponsored by the Japan International Co-operation Agency (JICA).

One representative pointed out the need to have a chain of training centres, including institutions each specializing in a particular field, in preference to a single training centre for the whole region. He also pointed out that high-resolution spot and LANDSAT TM data were expected to be more useful than LANDSAT MSS data for cartographic application.

The Committee recognized the need to start an ESCAP data bank on remote sensing to be included in the regional information system and reocmmended that information be disseminated through the *Remote Sensing Newsletter,* in view of the fact that a remote sensing journal might require greater effort and financial support.

The Committee noted with appreciation the offer of the Government of India to make available the data from the Indian remote sensing satellite which would be launched in 1988, and to share its experience in space through training programmes.

The Committee noted with appreciation the offer made by the USSR to share its expertise and experience in the use of space imagery in the production of geotectonic maps for various subregions of Asia and the Pacific. The Committee also noted that the USSR was considering a proposal to host in 1989 a seminar for the countries of the ESCAP region on the methodology used in making the maps.

The Committee was informed about the Geological Applications for Remote Sensing (GARS) programme of the International Union of Geological Sciences dealing with the dissemination of the results of remote sensing applications and techniques. GARS was currently active in Africa and could be expanded to the ESCAP region if requested.

G. Activities of other international bodies in the appraisal, development and management of energy resources
(Item 8 of the agenda)

The Committee took note of the papers circulated by the Department of Technical Co-operation for Development of United Nations Headquarters, the Economic Commission for Europe (ECE), the Energy Research Group of the International Development Research Centre (ERG/IDRC), FAO, the International Energy Agency (IEA), ILO, the Statistical

Institute for Asia and the Pacific (SIAP), the South Pacific Bureau for Economic Co-operation (SPEC), the World Energy Conference (WEC), and the World Bank concerning their involvement and activities in the field of energy. It also noted the information provided by the following organizations and United Nations bodies which made brief presentations to the Committee, reported below in the order of their presentations.

The International Union of Geological Sciences (represented by Norway) informed the Committee that IUGS and UNESCO had jointly organized a deposit modelling programme under which workshops had been held in Brazil, Chile and the Philippines. IUGS was prepared to organize similar workshops in the ESCAP region if requested to do so.

The Committee was informed that the Circum-Pacific Council, IUGS, CCOP and the Circum-Pacific Map Project (CPMP) had begun the preparation of a series of thematic maps at a scale of 1:2,000,000 for the joint resources assessment programme. The base map series consisted of eight sheets compiled by the United States Geological Survey, showing bathymetry, topography, drainage, and place names. The Committee expressed its appreciation to the United States for presenting a set of the maps to the secretariat for use in the compilation of information on the geology and natural resources of East Asia.

On the subject of the CPC East Asia Map Project, the Circum-Pacific Council (represented by the United States) informed the Committee that the resource assessment programme would include thematic maps covering several aspects of natural resources, and urged that the secretariat co-ordinate its activities with IUGS, CPMP, CCOP, the ASEAN Council on Petroleum (ASCOPE) and the International Geological Correlation Programme, all of which were involved in various aspects of natural resources assessment.

The representative of the Asian Institute of Technology gave an overview of its work in energy, human resources development and remote sensing, emphasizing post-graduate programmes in energy planning and economics, and rational use of energy; new and renewable sources of energy; and, finally, an interdisciplinary management of natural resources management programme using remote sensing technology.

The representative of the Department of International Economic and Social Affairs outlined the work done by the New and Renewable Sources of Energy Unit and the Energy Statistics Office, emphasizing the availability of a computerized data-base on multilateral and bilateral activities in new and renewable sources of energy that was updated regularly and available to interested users. He also informed the Committee on the preparations for the fourth session of the Intergovernmental Committee on New and Renewable Sources of Energy to be convened in March/April 1988.

The representative of the World Bank described its funding and technical assistance activities in energy resources development and in the field of efficient use of energy, which included: (i) loans with technical assistance, in primary energy resources development; (ii) the work of the project preparations office on energy issues (including pricing, marginal costing, and so on); (iii) the work of the UNDP/World Bank Energy Sector Assessment and Management Assistance Programme (ESMAP); and (iv) the work of the Economic Development Institute (EDI), offering training (mainly in the power sector).

The representative of the United Nations Environmental Programme described the work done in assessing the environmental impact of energy systems at the local, regional and global levels, recommending environmentally-sound approaches to all sources of energy. Although several models were available for such assessment, it was regrettable that those were not used to the extent advisable. He mentioned that a conference on energy-efficient environmental strategies for Thailand would be held at Pattaya from 4 to 6 March 1988; he informed the Committee about UNEP work in water resources management.

The representative of the Department of Technical Co-operation and Development gave an indication of the scope of technical assistance executed by the Department in energy projects in Bangladesh, China, India, the Philippines and Viet Nam. He emphasized the role played by the Department in propagating micro-computer-based energy planning technologies, including participation in an REDP activity by conducting a seminar on the use of ENERPLAN, a specifically developed software package available in the Department for "non-commercial" prospective users. He also emphasized experience gained in Latin America on a uniform system of energy balances showing net energy use, and informed the Committee that tools and training for use and interpretation of such systems were also available from the Department.

The representative of the United Nations Development Programme reported that UNDP attached considerable importance to the energy sector, as evidenced by its support to REDP and PEDP, executed by ESCAP, with a total investment of approximately $US 14 million. Apart from its contribution to the UNDP/World Bank ESMAP, worth $US 24 million, UNDP also supported 183 energy-related projects on a global basis, with a total UNDP investment of $US 111 million.

The Committee noted the above activities with interest.

H. Programme changes, 1988-1989, and proposed draft medium-term plan, 1990-1995
(Item 9 of the agenda)

1. Energy resources

The Committee had before it document E/ESCAP/ NR.14/11, entitled "Draft medium-term plan, 1990-1995, and programme changes, 1988-1989: energy resources".

The Committee was informed that the medium-term plan formulation process drew on all the policy studies presented to the Committee and that the draft plan presented had been discussed and generally endorsed by a group of senior experts prior to its submission to the Committee, while the general thrust of the plan had already been accepted by the Commission at its forty-third session in April 1987.

The Committee appreciated the presentation of the draft medium-term plan, 1990-1995, and agreed with the approach for meeting the felt regional needs. It urged the secretariat to undertake the necessary steps, including mobilizing adequate finance, to implement the plan.

Some delegations reiterated their earlier view that the regional energy development programme should continue after 1991.

The representative of China expressed the readiness of his country to contribute to the formation of the co-operative network on rural energy planning.

One delegation, while giving its general endorsement to the draft medium-term plan, expressed concern that certain parts of the report were not very clear. It stressed the importance of the training aspects of energy planning and demand management under subprogramme 1, commended the work plan for new and renewable sources of energy under subprogramme 2 and expressed readiness to support the proposed activities in wind-powered water pumping. Under subprogramme 3, concerning largely conventional sources of energy, the delegation supported the strengthening of TCDC activities. The delegation also suggested a link between the Asian and Pacific Energy Planning Network (APENPLAN) of APDC and the UNESCO Co-operative International Network for Training and Research in Energy Planning (CINTREP).

The Committee noted that there was no change proposed for the work programme, 1988-1989.

2. Water resources

In the water resources subprogramme, the Committee reviewed the draft medium-term plan, 1990-1995, and programme changes, 1988-1989, as set out in documents E/ESCAP/NR.14/22 and Add.1. The Committee noted that no changes were proposed for the programme of work, 1988-1989 in the document under the subprogramme concerning water resources.

The Committee endorsed the proposed draft medium-term plan, 1990-1995. While supporting the draft plan for the water sector, one delegation expressed concern at the time-limited subsidiary objective of the secretariat to establish an independent intergovernmental body by 1992 to take over from ESCAP the functions of the Regional Network for Training in Water Resources Development established in May 1986, as it might be too early to plan for such an objective.

The Committee endorsed the secretariat's initiative in presenting the subject of urban flood loss prevention and mitigation as a major issue to be submitted to the Commission for consideration at its forty-fourth session.

The representative of the USSR informed the Committee that his country would host a workshop on the role of water-use statistics in the long-term planning of water resources development scheduled to be held at Kiev in August 1988. He also confirmed USSR support for the workshop on management of the environmental impacts of water resources development projects.

In compliance with the request of the Commission at its forty-third session, the Committee considered the subject of shared water resources at length, but was unable to reach a conclusion. One delegation made the proposal that the secretariat prepare a general report on shared water resources development in the region, including case studies of successful implementation of shared water resources development, but excluding any bilateral problems, to be included in the programme of work, 1988-1989, and presented at the fifteenth session of the Committee. One delegation strongly opposed that proposal and said that it was clear that there was no consensus on that question. It also noted that the preparation of a study on shared water resources had been discussed at length in one form or another since the thirty-ninth session of the Commission held over four years ago and no consensus had been possible. In these circumstances, it recommended that that activity be dropped.

There was no conclusion in the absence of consensus. It was suggested that the subject be referred back to the Commission, at its forty-fourth session. It was also suggested that the proposal should not be further pursued in view of the clear absence of consensus.

One delegation pointed out that the relevant rule of procedure of ESCAP clearly indicated that it was not a must that every decision of the Commission and its committees should be on the basis of consensus. Three delegations stressed that the principle of consensus, which had guided all deliberations in ESCAP, was fundamental to the functioning of ESCAP and should continue to guide its deliberations.

Since there was neither conclusion nor consensus on the subject of shared water resources, it was suggested that the Commission be requested for further guidance on that subject at its forty-fourth session.

3. Mineral resources

In the mineral resources subprogramme, the Committee considered the proposed draft medium-term plan, 1990-1995, on the programme on natural resources as set out in document E/ESCAP/NR.14/22. The Committee was informed that in the medium-term plan, the Atlas of Stratigraphy and the project on stratigraphic correlation between sedimentary basins of the ESCAP region would be reoriented to emphasize the collation of published and other available information in a form that would permit effective correlation. The Committee recommended that the study of the region's sedimentary basins focus on their natural resources, especially on hydrocarbon, ground water, coal and non-metallic minerals, and that those studies be coordinated with other ongoing sedimentary basin resource assessment programmes.

The Committee also noted that the Fifth Working Group on Stratigraphic Correlation between Sedimentary Basins of the ESCAP Region and Atlas of Stratigraphy (Triassic) would meet at Bangkok in November 1987 and that the report of that meeting should form an additional basis for formulating future directions and activities relating to stratigraphic studies in the region. The Committee endorsed the medium-term plan in the mineral sector, noting that it would include urban geology and land-use planning as well as programmes on metallic and non-metallic minerals assessment, all of which were highly recommended.

The Committee also noted the proposal to continue the work on the Atlas of Mineral Resources until its logical completion in the form of a publication.

4. Marine resources

The Committee reviewed the proposed draft medium-term plan, 1990-1995 in marine affairs as set out in document E/ESCAP/NR.14/19.

The Committee expressed its support of the draft plan on marine affairs. It was recommended that that plan should be implemented by the secretariat in an integrated and multi-disciplinary fashion, in line with the United Nations Convention on the Law of the Sea as well as practices at the national level.

The Committee approved the strategy outlined in the medium-term plan in that it would assist in strengthening national capabilities in the assessment, development and management of the marine resources within national jurisdiction.

5. Cartography and remote sensing

With regard to the medium-term plan for the cartography and remote sensing subprogramme, as contained in document E/ESCAP/NR.14/22, one delegation objected to the establishment of an independent intergovernmental body, stating that sufficient preparatory work had not been done to justify the establishment of such a body and expressing concern about the financial viability of such a body. Another delegation proposed that it could be formed on a TCDC basis.

Another delegation sought clarification on the Professional post in the ESCAP secretariat that would be required if the regional remote sensing programme were to be integrated into the secretariat. In response, the Committee was informed that should that transpire, a post would be needed to serve the subprogramme as currently there was no full-time Professional assigned to that subprogramme. The Committee was also informed that that would not necessarily require the assignment of a new post and that the post could be filled by the redeployment of existing posts in the secretariat.

I. Consideration of the agenda and arrangements for subsequent sessions of the committee
(Item 10 of the agenda)

The Committee was informed that the draft provisional agenda for the fifteenth session of the Committee on Natural Resources, which would deal mainly with minerals and marine affairs, was presented in a modified format, through which policy options and human resources development in the mineral sector as well as assessment of geology for planning purposes, together with a review and analysis of the adjustments of national policies to the new regime of the sea, would form the core of the deliberations at the next session.

The Committee was also informed that in the light of the ongoing efforts within the United Nations to review the conference structure, it might be possible that the draft provisional agenda would have to be changed in the future. The draft provisional agenda for the sixteenth session, as adopted at the thirteenth session of the Committee in 1986, might also have to be changed at a later date.

The Committee attached great importance to industrial and non-metallic minerals and to the triennial review of mineral development in the ESCAP region, and those items, as well as a progress report on the stratigraphic correlation project, should be discussed at the fifteenth session of the Committee on Natural Resources.

The Committee also requested that a report on the use of cartography and remote sensing for resource evaluation and environmental protection be included in the draft provisional agenda.

With those additions, the Committee endorsed the draft provisional agenda for the fifteenth session.

J. Other matters
(Item 11 of the agenda)

No other matter was raised under that agenda item.

K. Adoption of the report
(Item 12 of the agenda)

On 2 November 1987, the Committee adopted the report on its fourteenth session, for consideration by the Commission at its forty-fourth session.

II. INFORMATION AND ISSUE PAPERS ON ENERGY

A. Current energy situation and demand management achievements

1. The regional energy scene*
(E/ESCAP/NR.14/9)

Introduction

The purpose of this paper is to highlight the current energy situation in the ESCAP region in respect of resources, production, consumption and demand based on the latest data available from the United Nations Statistical Office, United Nations and other publications and reports, and information collected from countries of the ESCAP region.

Although the statistical data used in this paper are not up to date because of the unavailability of the latest data, it is believed that substantial analyses and basic conclusions will indicate general trends in the region. This paper is mainly an information paper: issues are discussed in separate papers before the Committee.

---

* Note by the ESCAP secretariat.

(a) The scene in energy resources

Table 1 shows the proved reserves of fossil fuels as of the end of 1985. In terms of resources, coal reserves are the highest, followed by oil and natural gas, in the world. In the ESCAP region, the gas reserves are higher than those of oil. In respect of the reserve/production ratio, oil is the fastest depleting resource both in the world and in the region.

In the ESCAP region, nuclear energy is used for electricity generation only in India, Japan, Pakistan and the Republic of Korea. The resource situation in the ESCAP region compared with that in the world is shown in table 2.

Hydropower potential in the ESCAP region is the highest among all the regions in the world. On the other hand, it is the least exploited resource, compared with other regions. Table 3, although it does not cover the whole region, gives an overview of the situation in selected countries of the ESCAP region in terms of both potential and exploitation. For the first time the table has also included, wherever available, the small, mini and micro hydropower resources.

Table 1. Proved reserves of fossil fuels, end 1985
(Billion tons of coal equivalent)

	Oil		Natural gas		Coal	
	Amount	R/P ratio	Amount	R/P ratio	Amount	R/P ratio
World	143.70	34.4	132.73	57.6	954.5	219
Asia and Australia (excluding China and the Middle East)	3.75	15.9	6.36	50.9	85.4	230
China	3.60	19.4	1.15	66.5	99.0	117
Islamic Republic of Iran	9.75	59.1	17.98	100.0	–	–

Source: BP Statistical Review of World Energy, June 1986.
Notes: Proved reserves are generally taken to be those quantities which geological and engineering information indicates with reasonable certainty can be recovered in the future from known reservoirs under existing economic and operating conditions.
Reserves/production ratio: If the reserves remaining at the end of any year are divided by the production in that year, the result is the length of time that those remaining reserves would last if production were to continue at the then current level.

Table 2. Nuclear resources, 1983
(Uranium, metric tons)

	Reasonably assured	Estimated additional
World	1 455 800	896 800
ESCAP region	353 700	369 900
Australia	314 000	369 000
India	32 000	900
Japan	7 700	–

Source: 1984 Energy Statistics Yearbook (United Nations publication, Sales No. E/F.86.XVII.2)

Table 3. Water power potential and development in selected countries of the ESCAP region, 1984[a]

Country or area	Estimated potential of hydropower stations capacity[b] (MW)			Present installed capacity (MW)			Ratio of exploited capacity to estimated potential (percentage)	Basis or method of estimation
	Total	Medium and large	Small/mini and micro	Total	Medium and large	Small/mini and micro		
Bangladesh	451.3	450.3	1.3	130.0	130.0	..	28.8	Based on pre-feasibility studies
China	680 000.0	660 000.0	20 000.0	25 600.0	16 930.0	8 670.0	3.8	Q average flow
India	89 247.0	89 247.0[c]	—	13 858.0	13 737.0	121.0	15.5	90 per cent dependable water availability (Q 90); 60 per cent load factor
Indonesia	79 140.0	77 800.0	1 340.0	1 454.4	1 403.3	45.9 / 4.6 / 0.6	1.8	The potential was estimated on plant factor; run-of-the-river plant factor = 0.6 to 0.8; reservoir plant factor = 0.5
Islamic Republic of Iran	14 000.0	..	..	1 804.0	..	..	12.9	
Japan[d]	49 026.0	38 824.0	9 763.0 / 438.0 / 1.0	25 030.0	21 934.0	2 925.0 / 170.0 / 1.0	51.1	Economic potential estimated on the basis of (a) Run-of-the-river power plants with 50 per cent load factor; (b) Regulating pond power plants with a load factor of 40 per cent; (c) Reservoir power plants with a load factor of 20 per cent
Malaysia Peninsular Malaysia	3 032.2	3 008.0	24.2	1 236.2	1 234.0	2.2	40.8	Q average flow for hydropower stations above 10 MW installed capacity; Q 95 flow for hydropower stations below 10 MW installed capacity
Sabah[e]	1 900.0	..	..	66.3	66.0	0.3	3.5	At 90 per cent load factor
Sarawak	20 035.5	20 000.0	35.5	1.0	..	1.0	..	Technical potential and 0.5 plant factor (for medium and large hydropower stations)
New Zealand	5 710.0	..	..	4 427.0	4 300.0	127.0	77.5	Based on low flows (Q 95 and above) and sites within 10 km of inhabited areas (for small, mini and micro hydropower plants)
Pakistan	30 600.0	30 000.0	600.0	2 547.0	2 542.0	5.0	8.3	Available heads and average flow of water
Papua New Guinea	15 055.0	..	..	117.0	106.0	11.0	0.8	80 per cent load factor
Philippines	5 583.0	5 583.0	..	1 666.1	1 630.0	36.1	30.0	Based on the evaluation of 32 identified sites. Estimates of water power potential are based on reconnaissance and feasibility studies undertaken
Republic of Korea	3 012.0	..	..	1 202.0	1 192.0	10.0	39.9	Benefit/cost ratio method; for mini-hydro: Q 95 flow
Samoa	30.0	—	30.0	4.6	—	4.6	15.3	Average flow
Sri Lanka	2 000.0	..	..	542.0	536.0	6.0	27.1	Economic power potential

Country or area	Estimated potential of hydropower stations capacity[b] (MW)			Present installed capacity (MW)			Ratio of exploited capacity to estimated potential (percentage)	Basis or method of estimation
	Total	Medium and large	Small/mini and micro	Total	Medium and large	Small/mini and micro		
Thailand	8 009.1 (+14 045.0[f])	7 952.2	56.9	1 714.0	1 675.0	39.0	21.4	Q average flow
Vanuatu	57.0	–	57.0	–	–	–	–	

Source: Electric Power in Asia and the Pacific, 1983 and 1984 (United Nations publication, Sales No. E.86.II.F.23).

Notes: a Countries for which data were available.

  b Classification of hydropower stations: Up to 100 kW = micro hydro; 100 kW to 1,000 kW = mini-hydro; 1.0 MW to 10.0 MW = small hydro: above 10 MW = medium and large hydropower stations.

  c Including potential for small hydropower stations.

  d Pumped storage hydropower stations are not included.

  e Including 110 MW in Temon Pangi.

  f Potential of major tributaries of the Lower Mekong Basin.

(b) *The regional scene in energy supply*

(i) Primary energy

Included in the production of commercial primary energy for solids are hard coal, lignite, peat and oil shale; liquids comprise crude petroleum and natural gas liquids; gas comprises natural gas; and electricity comprises primary electricity generation from hydro, nuclear and geothermal resources.

The comparison between different fuels (except fuel-wood and bagasse) is presented in tons of coal equivalent (TCE) on the basis of heat energy that may be obtained under ideal conditions. In the case of primary electricity, the ideal condition is taken to be 3,412 British thermal units (BTU) per kilowatt-hour (kWh), which results in a coal equivalency of 0.123 tons per 1,000 kWh.

Table 4 shows the production of commercial energy in the ESCAP region in all forms: solids, liquids, gas and electricity. The production of primary energy as a whole in the region declined from 1,484.8 million TCE in 1978 to the lowest level of 1,224.3 million TCE in 1980, gradually recovered to 1,454.5 million TCE in 1983 and then increased significantly to 1,548.9 million TCE (6.5 per cent over the previous year) in 1984.

a. Solid fuels

As shown in table 4, except for a slight fall in 1980, the regional total production of solid fuels increased at an average annual growth rate of over 5.4 per cent during the period 1973-1984 compared with the overall commercial energy growth rate of only 2.3 per cent during the same period. Consequently, the share of solid fuels production in the regional total production has increased continuously, from 37.9 per cent in 1973 to 52.6 per cent in 1984, exceeding the share of liquid fuel production since 1979.

Table 5 shows the production of solid fuels in the ESCAP region distributed over major coal-producing countries. Australia, China and India, the three largest producers of solid fuels in the region, accounted for almost 95 per cent of the regional total solids production in 1984. China remained the largest producer, maintaining a share of between 66 and 70 per cent since 1973. Therefore, trends of regional solids production depend largely on the trend in China; the fall in regional production in 1980 was almost entirely due to the fall in production in that country. While Australia has been maintaining its share of between 11 and 12 per cent, the share of India's solids production has been steadily increasing, from 12 per cent in 1973 to 15 per cent in 1984. The share of other countries in the regional total production, although showing a slight increase in absolute terms, has been falling continuously, from about 9 per cent in 1973 to only 5 per cent in 1984.

b. Liquid fuels

The share of liquid fuel production has been following the opposite path to that of solid fuels. It has fallen from over 55 per cent in 1973 to 50 per cent in 1978, and further to 35 per cent in 1984 (table 4). As shown in table 6, as in the case of solid fuels, the production of liquid fuels is also centred around the three largest oil-producing countries, China, Indonesia and the Islamic Republic of Iran. After reaching a peak of 738 million TCE in 1978, the production of liquid fuels in the region fell suddenly to 596 million TCE in 1979, fell further to its lowest level of 453 million TCE in 1981 and then recovered slowly to 551 million TCE in 1984. This sudden fall may be attributed almost exclusively to the variation in the production of oil in the Islamic Republic of Iran, as evidenced from figures in table 6. The overall changes in production in other countries although falling slightly during 1980-1982, were not that significant. The share of

Table 4. Production of commercial energy in the ESCAP region
(Million tons of coal equivalent)

	1973	1978	1979	1980	1981	1982	1983	1984
Solids	456.426	617.259	632.314	625.378	661.300	703.369	747.693	814.840
	(37.92)	(41.57)	(46.12)	(51.08)	(52.66)	(51.36)	(51.4)	(52.64)
Liquids	665.758	737.850	596.391	460.596	453.199	517.457	540.980	550.528
	(55.32)	(49.69)	(43.5)	(37.62)	(36.09)	(37.79)	(37.19)	(35.56)
Gas	57.041	93.983	102.602	95.536	95.373	100.609	113.217	127.994
	(4.74)	(6.33)	(7.48)	(7.8)	(7.59)	(7.35)	(7.78)	(8.27)
Electricity	24.272	35.760	39.588	42.832	45.976	47.948	52.627	54.705
	(2.02)	(2.41)	(2.89)	(3.5)	(3.66)	(3.5)	(3.62)	(3.53)
Total	1 203.497	1 484.852	1 370.895	1 224.342	1 255.848	1 369.383	1 454.517	1 548.067
	(100)	(100)	(100)	(100)	(100)	(100)	(100)	(100)

Sources: United Nations, *Yearbook of World Energy Statistics*, various issues; and *1984 Energy Statistics Yearbook* (United Nations publication, Sales No. E/F.86.XVII.2).

Note: The figures in parentheses show the share as the percentage of total.

Table 5. Production of solid fuels in the ESCAP region
(Million tons of coal equivalent)

Country or area	1973	1978	1979	1980	1981	1982	1983	1984
Australia	52.770 (11.56)	68.397 (11.08)	71.551 (11.32)	69.927 (11.18)	79.511 (12.02)	84.242 (11.98)	89.928 (12.03)	95.063 (11.67)
China	307.143 (67.29)	433.139 (70.17)	445.436 (70.44)	434.954 (69.55)	436.327 (65.98)	467.728 (66.50)	501.521 (67.07)	553.821 (67.97)
India	56.717 (12.43)	73.722 (11.94)	74.967 (11.86)	79.431 (12.70)	103.828 (15.70)	108.677 (15.45)	114.101 (15.26)	122.559 (15.04)
Others	39.796 (8.72)	42.001 (6.81)	40.360 (6.38)	41.066 (6.57)	41.634 (6.30)	42.722 (6.07)	42.143 (5.64)	43.397 (5.32)
ESCAP total	456.426 (100)	617.259 (100)	632.314 (100)	625.378 (100)	661.300 (100)	703.369 (100)	747.693 (100)	814.840 (100)

Sources: United Nations, *Yearbook of World Energy Statistics,* various issues; and *1984 Energy Statistics Yearbook* (United Nations publication, Sales No. E/F.86.XVII.2).

Note: The figures in parentheses show the share as a percentage of the ESCAP total.

the Islamic Republic of Iran in the regional liquid fuel production dropped from 64 per cent in 1973 to 52 per cent in 1978 and 23 per cent in 1980 and 1981, though recovering later to over 30 per cent after 1982. In 1984, its share was 29 per cent of the total production for the region. In that year, the production in China again surpassed that of the Islamic Republic of Iran, as in 1980 and 1981. Since its first decline in 1982, the production in Indonesia remained low, with a tendency to recover slowly in the latter years. Australia maintained its production at more or less the same level, although in 1984 production showed a sharp rise (15.8 per cent) over the previous year.

c. Gaseous fuels

Referring to table 4, it may be seen that regional gas production has continued to grow almost steadily over the last decade, raising its share from 4.7 per cent in 1973 to 8.3 per cent in 1984. The production grew at an average annual rate of 7.6 per cent. Natural gas is found more evenly in the region than coal and oil. Table 7 shows the production of gaseous fuels in the countries of the region. It may be observed from the table that because of increased production in countries with smaller quantities, the production grouped under "others" has increased significantly in both absolute terms and in terms of regional share. Except in the Islamic Republic of Iran, where the production fell sharply in 1980 and remained low, the production of gaseous fuels in other countries of the region was quite high. However, in Brunei Darussalam production has remained at almost the same level since 1978. In China, it fell slightly in 1981 and has remained virtually unchanged since then.

d. Primary electricity

As shown in table 4, the production of primary electricity increased steadily during the years 1973-1984 at an average annual rate of 7.7 per cent. Although its share in

the regional energy supply is small, in some countries it play an important role in overall electricity supply. The regional electricity production, including secondary electricity, is analysed in the following subsection.

(ii) Electricity

The present review covers only the 24 countries or areas which contributed to the biennial publication *Electric Power in Asia and the Pacific 1983 and 1984*. Regional totals on installed generating capacity and electricity generation therefore exclude all other countries of the ESCAP region. Data do not include the capacity and generation of self-generating industries (table 8).

a. Utility size and structure

Owing to considerable disparities in the size of countries and degrees of economic development, installed generating capacities and power generation vary greatly throughout the ESCAP region. Installed generating capacity ranges from 2.38 MW in Kiribati (1984) to 148,337 MW in Japan (1984). Power generation during the same year was 5.5 GWh in Kiribati compared with 582,195 GWh in Japan.

Only four countries in the ESCAP region had nuclear power plants for electricity generation: India, Japan, Pakistan and the Republic of Korea. During the period 1981-1984, the installed capacity in nuclear power plants increased from 17,649 MW in 1981 by 34.3 per cent to 23,709 MW in 1984.

Geothermal power is used for electricity generation in China, Indonesia, Japan, New Zealand and the Philippines. Since 1981, installed capacity in public electric utilities increased from 772 MW by 64.2 per cent to 1,268 MW in 1984.

Table 6. Production of liquid fuels in the ESCAP region
(Million tons of coal equivalent)

Country or area	1973	1978	1979	1980	1981	1982	1983	1984
Australia	29.321	31.750	31.425	30.009	30.396	30.343	29.709	34.416
	(4.41)	(4.30)	(5.27)	(6.52)	(6.71)	(5.86)	(5.49)	(6.25)
China	74.518	151.289	154.342	154.051	147.176	148.487	154.223	166.647
	(11.19)	(20.50)	(25.88)	(33.44)	(32.47)	(28.70)	(28.51)	(30.27)
India	10.466	16.388	18.671	13.666	21.726	28.769	36.635	40.744
	(1.57)	(2.22)	(3.13)	(2.97)	(4.79)	(5.56)	(6.77)	(7.40)
Indonesia	97.354	116.711	113.532	112.891	114.654	95.706	96.158	99.633
	(14.62)	(15.82)	(19.04)	(24.51)	(25.30)	(18.50)	(17.78)	(18.10)
Islamic Republic of Iran	428.128	385.207	234.566	106.814	105.720	176.214	179.983	159.682
	(64.31)	(52.21)	(39.33)	(23.19)	(23.33)	(34.05)	(33.27)	(29.01)
Others	25.971	36.505	43.865	43.165	33.527	37.938	44.272	49.406
	(3.90)	(4.95)	(7.35)	(9.37)	(7.40)	(7.33)	(8.18)	(8.97)
ESCAP region total	665.758	737.850	596.391	460.596	453.199	517.457	540.980	550.528
	(100)	(100)	(100)	(100)	(100)	(100)	(100)	(100)
ESCAP region total (excluding Islamic Republic of Iran)	237.630	352.643	361.835	353.782	347.479	341.243	360.997	390.846

Sources: United Nations, *Yearbook of World Energy Statistics,* various issues; and *1984 Energy Statistics Yearbook* (United Nations publication, Sales No. E/F.86.XVII.2).
Note: Figures in parentheses show the share as a percentage of the total for the ESCAP region.

Table 7. Production of gaseous fuels in the ESCAP region
(Million tons of coal equivalent)

Country or area	1973	1978	1979	1980	1981	1982	1983	1984
Australia	4.498	8.596	9.602	11.140	14.124	15.628	15.809	16.681
	(7.89)	(9.15)	(9.36)	(11.66)	(14.81)	(15.53)	(13.97)	(13.03)
Brunei Darussalam	2.613	11.982	12.857	13.617	11.432	11.420	11.584	12.028
	(4.58)	(12.75)	(12.53)	(14.25)	(11.99)	(11.35)	(10.23)	(9.40)
China	9.417	18.261	19.299	19.039	16.917	15.841	16.213	16.505
	(16.51)	(19.43)	(18.81)	(19.93)	(17.74)	(15.75)	(14.32)	(12.90)
Indonesia	2.180	12.059	16.444	20.725	22.252	22.054	26.380	32.090
	(3.82)	(12.83)	(16.03)	(21.69)	(23.33)	(21.92)	(23.30)	(25.07)
Islamic Republic of Iran	24.128	22.795	23.870	9.506	8.421	8.757	11.121	11.164
	(42.30)	(24.25)	(23.26)	(9.95)	(8.83)	(8.70)	(9.82)	(8.72)
Pakistan	4.631	6.505	7.021	8.395	9.292	10.020	10.549	10.566
	(8.12)	(6.92)	(6.84)	(8.79)	(9.74)	(9.96)	(9.32)	(8.26)
Others	9.574	13.785	13.509	13.114	12.935	16.889	21.561	28.960
	(16.78)	(14.67)	(13.17)	(13.73)	(13.56)	(16.79)	(19.04)	(22.62)
ESCAP region total	57.041	93.983	102.602	95.536	95.373	100.609	113.217	127.994
	(100)	(100)	(100)	(100)	(100)	(100)	(100)	(100)

Sources: United Nations, *Yearbook of World Energy Statistics,* various issues; and *1984 Energy Statistics Yearbook* (United Nations publication, Sales No. E/F.86.XVII.2).
Note: Figures in parentheses show the share as percentage of the total for the ESCAP region.

Table 8. Installed electricity generating capacity and electricity generation
(by category) of electric power utilities in the
Asian and Pacific countries, 1984[a]

Country or area	Installed capacity (MW)	Generation (GWh)
Category I (capacity up to 10 MW)		
Kiribati	2.38	5.5
Maldives	3.50	8.6
Vanuatu	6.37	17.3
Category II (capacity from 10 to 100 MW)		
Bhutan	10.0	11.6
Samoa	16.3	33.8
Category III (capacity from 100 to 1,000 MW)		
Burma	636.0	1 804.0
Fiji	161.8	306.8
Papua New Guinea	201.0	433.0
Sri Lanka	812.0	2 261.0
Category IV (1,000 to 10,000 MW)		
Bangladesh	1 121.0	3 966.2
Hong Kong	5 265.3	17 918.0
Indonesia	4 515.1	13 621.5
Malaysia[b]	3 483.4	13 243.8
New Zealand	6 988.0	26 746.0
Pakistan	5 164.0	21 895.0
Philippines	5 458.4	19 132.9
Singapore	2 691.0	9 421.0
Thailand	6 128.0[d]	21 024.6
Category V (over 10,000 MW)		
Australia	31 230.0	103 362.3
China[c]	74 152.2	353 462.0
India	39 339.0	139 956.0
Islamic Republic of Iran	11 419.0	34 094.0
Japan	148 337.0	582 195.0
Republic of Korea	14 190.0	53 808.0

Source: Electric Power in Asia and the Pacific 1983 and 1984 (United Nations publication, Sales No. E.86.II.F.2.3)

Notes: [a] Data exclude capacity and generation of self-generating industries.

[b] Aggregate data of Peninsular Malaysia, Sabah and Sarawak.

[c] Excluding power stations below 500 kW capacity.

[d] Figure updated during the Committee Session.

For the purpose of comparison, electric power utility systems of ESCAP members have been grouped in table 8 as follows:

i.   Category I – Very small power systems (up to 10 MW)

Electricity in these island countries was produced exclusively by internal combustion engines and amounted to 5.5 GWh, 8.6 GWh and 17.3 GWh, respectively in Kiribati, Maldives and Vanuatu. Vanuatu has a hydropower potential of the order of 57 MW.

ii.   Category II – Small power systems (10 MW to 100 MW)

Of the total installed capacity in Bhutan of 10 MW (1984), 3.5 MW or 35 per cent was hydropower. In Samoa, the total installed capacity in 1984 was 16.3 MW, which generated 33.8 GWh, shared almost equally by hydropower and internal combustion engines.

iii.   Category III – Medium-sized power systems (100 to 1,000 MW)

All the countries under this category have a mixed system of hydro and thermal power, with hydro in 1984 contributing 55, 100, 32 and 92 per cent respectively in Burma, Fiji, Papua New Guinea and Sri Lanka.

iv. Category IV — Large power systems (1,000 to 10,000 MW)

The majority of the countries and areas considered by this review are in this category. With the exception of Hong Kong and Singapore, where electricity is produced exclusively from thermal power, all countries produce electricity in both hydropower and thermal power stations. In addition, Pakistan produces a small amount of electricity from nuclear power.

v. Category V — Very large power systems (over 10,000 MW)

With an installed capacity of 148,337 MW, the capacity of Japan is twice as high as that of China and almost four times that of India. The largest percentage of installed capacities in this bracket comes from steam power plants, ranging from 73 per cent in Australia to 48 per cent in the Islamic Republic of Iran. Hydropower is second in importance: China and India have approximately 35 per cent each of their total installed capacity in the hydropower sector. In the Islamic Republic of Iran, 28.6 per cent of the installed capacity is in gas turbine plants.

b. Generation

As a consequence of a steady increase in electric power generation in the region as a whole, the quantity of primary and secondary electricity produced almost reached the 1.5 million GWh mark in 1983 and exceeded it in 1984. Whereas in 1973/74 the annual increase in power generation had reached an unprecedented low of 0.48 per cent, it established itself at the 5.2 to 6.5 per cent level during subsequent years. From 1979 onwards, regional totals include data for China; therefore, the increase from 874,961.5 GWh in 1978 by 36.7 per cent to 1,195,834.1 in 1979 does not reflect the real situation in the region. Without the electricity generated by China, the increase in the region from 1978 to 1979 would have been 4.4 per cent, with only a minor increase of 0.46 per cent from 1979 to 1980, and widely varying increases ranging from 2.37 to 9.21 per cent during the period 1981-1984 (table 9).

An examination of developments in the generation of primary electricity, including hydropower, nuclear energy, geothermal energy and, from 1981 onwards, some other new and renewable sources on energy, reveals an almost uninterrupted upward trend in the region from 1973 onwards, ranging from 2.2 per cent (1971-1972) to 16.6 per cent (1980-1981).

Up to 1973, the yearly growth rate of secondary electricity generation was above 10 per cent; during the subsequent two years, power generation in thermal power stations declined, whereas primary electricity generation experienced a growth of 17.1 per cent in 1973/74 and 9 per cent in 1974/75. In 1976, generation of secondary electricity increased again, by 7.1 per cent over the level of the previous year and 6.4 per cent over the 1973 level.

The highest growth rate during the period 1976-1984 occurred in 1976-1977 with a 10.7 per cent increase, whereas for the remainder of the time generation increase ranged from 2.0 to 9.9 per cent.

c. Generation changes by types of plant

It may be noted that in 1971, generation of electricity from hydropower was 17 times higher than that from nuclear power stations. Owing to the steep rise in electricity generation from nuclear energy, mainly in Japan and later in the Republic of Korea, this ratio dropped to 2.6 in 1978 and to 1.3 in 1984. In 1984, geothermal power represented 1.7 per cent of total primary electricity generation, and other new and renewable sources of energy (excluding hydro), 0.25 per cent.

Throughout the period 1971-1984, steam occupied the first place in secondary electricity generation, with approximately 96 per cent. Over the period, generation by gas turbines increased. By 1984, gas turbines and internal combustion engines were generating almost equal shares of 2 per cent each.

In total electricity generation (primary and secondary), steam was the most important source of power generation, with 68.5 per cent in 1971, a peak of 76.2 per cent in 1973, and declining importance from 1979 onwards. The share of generation by steam dropped to a low of 49.9 per cent in 1982, but recovered to 66 per cent in 1984.

Next in importance is hydropower generation, which developed at a relatively steady rate from 1971 onwards and generated, in 1984, 17.4 per cent of the total, followed by nuclear energy with 12.9 per cent, gas turbines with 1.6 per cent, internal combustion engines with 1.5 per cent, geothermal with 0.5 per cent and new and renewable sources of energy with a negligible amount.

d. Development in generating capacity

It should be kept in mind that the countries making up regional yearly totals varied during the period under review. The installed generating capacity in the region (excluding China) increased between 1971 and 1984 by an annual average of 7.1 per cent. The highest yearly increase (11 per cent) took place between 1971 and 1972, and the least significant (3.9 per cent) between 1978 and 1979. Between 1980 and 1981, the yearly regional growth again reached 10.4 per cent.

The average annual growth rate of the developing countries of the region is exactly double (8.5 per cent) that of the developed countries (4.2 per cent).

e. Trends in transmission systems

As of 1984, high-voltage transmission lines (500 kV) were added to the grid in Australia, China (840 circuit km), Indonesia (238 circuit km), Japan (492 circuit km), and

Table 9. Trend in electricity production in the ESCAP region
(Gigawatt hours (GWh))

Types of generation / Year	1973	1974	1978	1979	1980	1981	1982	1983	1984
Primary electricity	157 173.0	184 014.2	222 174.6	297 789.1	321 971.7	375 475.5	391 365.1	427 322.2	445 829.0
Hydro	144 667.4	161 449.2	161 080.6	221 367.1	233 026.7	272 888.6	275 713.2	293 940.9	289 063.0
Nuclear	12 505.6	22 565.0	61 094.0	76 422.0	88 945.0	98 738.0	110 870.0	126 976.0	149 872.0
Geothermal						3 848.9	4 781.9	5 480.3	5 998.0
NRSE								925.0	896.0
Secondary electricity	544 842.1	521 385.2	652 786.9	898 128.2	894 446.8	929 060.8	952 090.2	1 033 092.0	1 096 237.4
(Thermal)									
Steam	534 558.9	505 699.8	629 691.4	873 896.9	870 389.2	902 895.3	923 302.7	994 955.1	1 059 000.5
Int. combustion	7 721.4	12 619.7	13 988.2	13 663.1	11 993.7	14 669.7	15 850.8	18 822.1	18 026.3
Gas turbines	2 561.8	3 065.7	9 107.3	10 568.2	12 063.9	11 495.8	12 936.7	19 314.8	19 210.6
Total*	702 015.1	705 399.4	874 961.5	1 195 834.1**	1 218 676.0**	1 304 536.3**	1 343 455.3**	1 460 414.2	1 542 066.4
Percentage increase in generation		0.48	36.70	1.90	7.70	3.20	7.85	5.59	
Total (excluding China)	702 015.1	705 399.4	874 961.5	913 834.1**	918 076.0**	1 002 672.6**	1 026 401.9**	1 108 975.2	1 165 075.4
Percentage increase in generation (excluding China)		0.48	4.44	0.46	9.21	2.37	8.04	5.06	

Source: United Nations, *Electric Power in Asia and the Pacific,* various issues.

*   Regional totals exclude data for those countries for which no data were available; from 1979 onwards, regional totals include data for China.

**   Including unclassified generation in some countries.

Pakistan (438 circuit km). In Hong Kong and India, 400 kV lines were in operation; they were doubled in the Islamic Republic of Iran to reach 4,500 circuit km. In Australia and China, the grids were expanded by 330 kV lines and in the Republic of Korea, the length of 345 kV lines was extended by one third.

(iii) Non-commercial energy

Although limited trading takes place for certain types of traditional sources of energy, they are still widely recognized as non-commercial sources of energy. Again, most of those are renewable. Renewable sources of energy fall broadly into three categories: biomass in traditional forms (wood and agricultural residues); biomass in non-traditional forms (converted into liquid or gaseous fuels); and solar, wind, mini-hydropower etc. Some information on hydropower, including mini- and micro-hydropower, have been given in table 3. A separate paper[1] deals with the contribution of new and renewable sources of energy to regional energy supply.

Non-commercial energy accounts for a large share of the energy supply in developing countries of the ESCAP region. In some countries, the share goes over 80 per cent of the total energy supplies. However, for lack of comparable and reliable data, it is hard to estimate the exact amount of such energy production in the region. An attempt has been made in tables 10 and 11 to give an overall picture in the world and in the ESCAP region. It may be noted here that most of the data on fuelwood are estimated by the United Nations Statistical Office.

(c) The regional scene in commercial energy demand

(i) Recent trends in regional energy consumption

---

[1] "Updated assessment of the contribution of new and renewable sources of energy to regional energy supply" (E/ESCAP/NR. 14/6).

Included in the consumption of commercial energy for "solid fuels" are consumption of primary forms of solid fuels, net imports and changes in stocks of secondary fuels; "liquid fuels" comprise consumption of energy petroleum products (including feedstocks), natural gasolene, condensate, refinery gas and input of crude petroleum to thermal power plants; gases include the consumption of natural gas, net imports and changes in stocks of gas works and coke-oven gas; and "electricity" comprises production of primary electricity and net imports of electricity.

As shown in table 12, the regional primary energy consumption increased at a rate of about 5 per cent per annum during the period 1973-1978 to 1,424 million TCE in 1978, remained almost at the same level until 1981 and then started increasing progressively at 3, 5 and 7 per cent in 1982, 1983 and 1984 respectively. Except in 1984, all increases were due to higher consumption in developing countries. While, during 1978-1983, the consumption in developed countries remained at almost the same level, it increased significantly by 8 per cent in 1984.

One significant change in the regional energy scene, observed by comparing the production and consumption of primary energy (tables 4 and 12), is that the region has turned into a net importer since 1979, when the consumption was 1,473 million TCE compared with a production of 1,371 million TCE. This is mainly due to the sharp fall of liquid fuel production in the region, particularly in the Islamic Republic of Iran.

Although the per capita energy consumption in developing countries had been improving gradually, it still remained low. In 1984, the average per capita commercial energy consumption in developing countries was 456 kg of coal equivalent, compared with the regional average of 643 kg and the world average of 1,859 kg. Of course, in many developing countries non-commercial energy constitutes a major share. Nevertheless, the overall energy consumption is still low.

Table 10. Production of fuelwood and bagasse, 1984

	Fuelwood[a] (Thousands of cubic metres)	Bagasse (Thousands of metric tons)
World	1 657 515	218 855
Africa	370 466	22 161
Asia	726 942	63 701
Europe	70 327	48
North America	156 141	51 616
Oceania	12 913	8 268
South America	240 426	73 061
USSR	80 300	

Source: *1984 Energy Statistics Yearbook*, (United Nations publication, Sales No. E/F.86.XVII.2).

Note: [a] Figures are mostly estimated by the United Nations Statistical Office.

Table 11. Production of fuelwood and bagasse in the ESCAP region[a]

	1981	1982	1983	1984
Fuelwood[b]				
(Thousands of cubic metres)	672 932	683 567	694 429	705 397
Bagasse				
(Thousands of metric tons)	60 610	77 627	72 620	69 806

Source: 1984 *Energy Statistics Yearbook,* (United Nations publication, Sales No. E/F.86.XVII.2).

Notes: [a] Over 99 per cent of the fuelwood and 80-90 per cent of bagasse production were in developing countries.

[b] Figures are mostly estimated by the United Nations Statistical Office.

Table 12. Consumption of commercial primary energy
(Million tons of coal equivalent and kilogrammes of coal equivalent per capita)

	1973	1978	1979	1980	1981	1982	1983	1984
World	7 438.087 (1 923)	8 369.716 (1 967)	8 653.468 (2 000)	8 566.208 (1 946)	8 457.419 (1 867)	8 428.058 (1 830)	8 549.740 (1 825)	8 855.792 (1 859)
ESCAP region	1 116.413 (527)	1 424.204 (605)	1 473.076 (615)	1 452.877 (586)	1 472.317 (585)	1 513.951 (591)	1 592.434 (611)	1 711.523 (643)
ESCAP developed countries	486.346 (3 889)	520.157 (3 933)	533.586 (4 015)	528.750 (3 926)	523.849 (3 852)	516.645 (3 771)	520.009 (3 795)	561.800 (4 042)
ESCAP developing countries	630.067 (314)	904.047 (402)	939.490 (410)	924.127 (398)	948.468 (399)	997.306 (411)	1 072.425 (435)	1 149.723 (456)

Source: Based on *Yearbook of World Energy Statistics,* various issues; 1984 *Energy Statistics Yearbook* (United Nations publication, Sales No. E/F.86.XVII.2) and *Statistical Yearbook for Asia and the Pacific 1984* (United Nations publication, Sales No. E/F.85.II.F.21).

Note: The figures in parentheses show per capita consumption (kg).

Table 13 has been prepared as a comparison between various groups of countries in the ESCAP region in terms of income level. The grouping has been chosen from the World Bank classification. In the lower-income countries, except China and India, the consumption of energy is very low. The consumption even in India is below the average for the developing countries of the region.

The consumption pattern of commercial primary energy in the ESCAP region is illustrated in table 14, and this is followed by similar illustrations with respect to developed and developing countries, in tables 15 and 16 respectively. The share of consumption of solid fuels increased to over 50 per cent in 1984, and that of liquid fell to 39 per cent. The consumption of gaseous fuel and primary electricity increased steadily from 1973 to 1984. One thing very clear from tables 14, 15 and 16 is the inter-fuel substitution from liquid fuels to other sources in both developed and in developing countries, although less prominently in the developing countries.

(ii) Projection of primary energy demand

Energy projections are basically estimations of likely energy demands and are generally made on somewhat

arbitrary assumptions of factors that influence energy demand: therefore, the analyses should be taken only as an indication. In a separate paper[2] before the Committee, a likely regional scenario has been worked out which was reviewed and commented upon by the Meeting of Senior Experts Preparatory to the Committee.[3] Projections will depend mainly on oil price assumptions. Another factor is the elasticity, which is the ratio of the percentage increase of energy consumption to that of GDP. Elasticity in most of the developing countries of the region is more than 1, with a few exceptions where the elasticity is close to 1. Therefore the demand for energy is likely to grow at a faster rate than GDP in developing countries unless strong conservation efforts are made to bring the elasticity down, such as those being pursued in countries like China and the Republic of Korea. The situation in developed countries is different, as the elasticity is low and conservation measures are widely practised.

[2] "Prospects for production and utilization of coal, natural gas and electricity" (E/ESCAP/NR.14/10).

[3] "Report of the Meeting of Senior Experts Preparatory to the Fourteenth Session of the Committee on Natural Resources" (E/ESCAP/NR.14/4).

Table 13. Consumption of commercial primary energy in developing countries of the ESCAP region
(Million tons of coal equivalent and kilogrammes of coal equivalent per capita)

Country or area	1981	1982	1983	1984
Low-income countries	739.287 (365)	778.007 (377)	837.689 (399)	902.732 (421)
China	555.917 (562)	583.321 (579)	628.499 (616)	684.009 (664)
India	148.907 (212)	157.564 (220)	170.545 (233)	176.904 (237)
Others	34.463 (107)	37.122 (112)	38.645 (114)	41.819 (121)
Lower-middle-income countries	74.59 (297)	77.498 (301)	83.254 (317)	84.225 (314)
Indonesia	38.783 (252)	40.901 (261)	44.625 (280)	42.579 (263)
Mongolia	2.62 (1 533)	2.904 (1 655)	2.973 (1 649)	3.015 (1 629)
Papua New Guinea	0.968 (291)	0.97 (284)	0.996 (284)	1.025 (285)
Philippines	15.67 (316)	15.649 (308)	15.726 (302)	16.684 (312)
Thailand[a]	17.819 (372)	18.237 (373)	20.189 (408)	22.060 (436)
Upper-middle and high-income countries/areas	134.585 (1 336)	141.801 (1 376)	151.482 (1 444)	162.766 (1 516)
Brunei Darussalam	2.952 (12 300)	3.318 (13 272)	2.662 (10 238)	2.999 (11 149)
Fiji	0.479 (747)	0.372 (571)	0.317 (478)	0.285 (423)
Hong Kong	7.921 (1 535)	8.535 (1 617)	9.463 (1 756)	9.681 (1 761)
Islamic Republic of Iran	41.441 (1 036)	46.62 (1 131)	49.423 (1 163)	58.154 (1 328)
Malaysia	12.067 (850)	12.704 (874)	12.845 (864)	13.309 (875)
Republic of Korea	55.934 (1 446)	54.991 (1 402)	58.466 (1 471)	61.424 (1 524)
Singapore	13.791 (5 640)	15.261 (6 164)	18.306 (7 299)	16.914 (6 659)
All developing countries of ESCAP region	948.468 (399)	997.306 (411)	1 072.425 (435)	1 149.723 (456)

Sources: 1984 Energy Statistics Yearbook (United Nations publication, Sales No. E/F.86.XVII.2); *Statistical Yearbook for Asia and the Pacific 1984* (United Nations publication, Sales No. E/F.85.II.F.21); World Bank, *World Development Report 1986.*

Note: The figures in parentheses show per capita consumption in kilogrammes.

[a] Figures updated during the Committee Session.

Table 14. Consumption pattern of commercial primary energy in the ESCAP region
(Million tons of coal equivalent)

	1973	1978	1979	1980	1981	1982	1983	1984
Solids	489.028	631.647	656.032	664.822	703.613	743.403	785.39	856.934
	(43.80)	(44.35)	(44.53)	(45.76)	(47.79)	(49.1)	(49.32)	(50.07)
Liquids	558.403	671.552	682.561	648.376	625.173	617.933	641.943	668.41
	(50.02)	(47.15)	(46.34)	(44.63)	(42.46)	(40.82)	(40.31)	(39.05)
Gas	44.711	85.23	94.862	96.804	97.509	104.579	112.434	131.475
	(4.00)	(5.98)	(6.44)	(6.66)	(6.62)	(6.91)	(7.06)	(7.68)
Electricity	24.271	35.775	39.621	42.875	46.021	48.039	52.665	54.7
	(2.17)	(2.51)	(2.69)	(2.95)	(3.13)	(3.17)	(3.31)	(3.2)
Total	1 116.413	1 424.204	1 473.076	1 452.877	1 472.316	1 513.954	1 592.432	1 711.519
	(100)	(100)	(100)	(100)	(100)	(100)	(100)	(100)

Sources: United Nations, *Yearbook of World Energy Statistics,* various issues; and *1984 Energy Statistics Yearbook* (United Nations publication, Sales No. E/F.86.XVII.2).

Note: The figures in parentheses show the percentage of the total.

Table 15. Consumption pattern of commercial primary energy in the developed countries of the ESCAP region
(Million tons of coal equivalent)

	1973	1978	1979	1980	1981	1982	1983	1984
Solid	107.939	101.963	108.885	120.864	135.165	135.718	130.780	144.590
	(22.19)	(19.60)	(20.33)	(22.85)	(25.80)	(26.27)	(25.15)	(25.74)
Liquids	352.828	363.151	363.811	336.271	312.919	300.870	304.320	317.781
	(72.55)	(69.82)	(67.93)	(63.60)	(59.73)	(58.24)	(58.52)	(56.56)
Gas	12.173	34.527	39.633	46.137	49.327	52.813	55.719	70.338
	(2.50)	(6.64)	(7.40)	(8.73)	(9.42)	(10.22)	(10.72)	(12.52)
Electricity	13.406	20.516	23.257	25.478	26.437	27.246	29.190	29.092
	(2.76)	(3.94)	(4.34)	(4.82)	(5.05)	(5.27)	(5.61)	(5.18)
Total	486.346	520.157	535.586	528.750	723.849	516.645	520.009	561.800
	(100)	(100)	(100)	(100)	(100)	(100)	(100)	(100)

Sources: United Nations, *Yearbook of World Energy Statistics,* various issues; and *1984 Energy Statistics Yearbook* (United Nations publication, Sales No. E/F.86.XVII.2).

Note: The figures in parentheses show the percentage of the total.

**Table 16. Consumption pattern of commercial primary energy in the
developing countries of the ESCAP region**

(Million tons of coal equivalent)

	1973	1978	1979	1980	1981	1982	1983	1984
Solid	381.089	529.684	547.147	543.958	568.448	607.685	654.610	712.344
	(60.48)	(58.59)	(58.36)	(58.87)	(59.94)	(60.93)	(61.04)	(61.96)
Liquids	205.575	308.401	318.750	312.105	312.254	317.063	337.623	350.629
	(32.63)	(34.11)	(34.00)	(33.77)	(32.92)	(31.79)	(31.48)	(30.50)
Gas	32.538	50.703	55.229	50.667	48.182	51.766	56.715	61.137
	(5.16)	(5.61)	(5.89)	(5.48)	(5.08)	(5.19)	(5.29)	(5.32)
Electricity	10.865	15.259	16.364	17.397	19.584	20.793	23.475	25.608
	(1.73)	(1.69)	(1.75)	(1.88)	(2.06)	(2.09)	(2.19)	(2.22)
Total	630.067	904.047	937.490	924.127	948.468	997.306	1 072.425	1 149.723
	(100)	(100)	(100)	(100)	(100)	(100)	(100)	(100)

Sources: United Nations, *Yearbook of World Energy Statistics*, various issues; and *1984 Energy Statistics Yearbook* (United Nations publication, Sales No. E/F.86.XVII.2).

Note: The figures in parentheses show the percentage of the total.

Table 17 shows the projected average annual growth rates of primary commercial energy demand in the developed countries and selected developing countries of the ESCAP region. The average projected growth rates in developed countries vary from 1 to 3 per cent, while those in developing countries range from 4 to as high as 9 per cent per annum.

The likely pattern of energy demand, broken down by sources, is shown in table 18. Compared with the projection reported to the Committee in 1984, the Republic of Korea and Thailand have revised the projected share of oil upwards, while New Zealand, a developed country, has lowered the share of oil demand.

**Table 17. Projected average annual growth rates of primary commercial energy
demand in selected countries in the ESCAP region**

(Percentage)

Country or area	Periods		
	1984-1990	*1990-2000*	*2000-2010*
Developed countries			
Australia	2.53	1.55	n.a.
Japan	0.96	2.77	n.a.
New Zealand	3.26	1.63	n.a.
Developing countries			
Bangladesh	8.7	n.a.	n.a.
China	4.38 (1985-2000)		3.95-5.28
Indonesia	3.68	5.07	4.46
Nepal	4.0 (1983/84-1990/91)	n.a.	n.a.
Republic of Korea	3.87-6.88 (1985-1991)	3.73-4.60 (1991-2001)	n.a.
Thailand[a]	6.81	4.48 (1990-2001)	n.a.
Viet Nam	5.42 (1983-2010)		

Sources: For developed countries: OECD/International Energy Agency, *Energy Policies and Programmes of IEA Countries, 1985 Review;* for developing countries: national sources.

Notes: n.a. indicates that projections are not available.

[a] Figures updated during the Committee Session.

Table 18. Projected energy demand pattern in selected countries in the ESCAP region
(Percentage)

Country	Period	Coal	Oil	Gas	Hydro	Nuclear	Others
Australia	1984	43.0	38.7	14.3	4.0	–	
	1990	45.6	33.8	16.4	4.1	–	
	2000	47.4	31.6	16.9	4.0	–	
Japan	1984	18.5	59.0	8.8	5.1	8.7	
	1990	18.0	52.0	11.9	6.4	11.7	
	2000	20.8	45.3	10.5	6.2	17.3	
New Zealand	1984	12.7	29.9	19.8	37.6	–	
	1990	17.4	21.6	26.1	34.9	–	
	2000	18.9	21.7	24.2	35.2	–	
China	1985	75.9	17.1	2.3	4.8	–	–
	2000	67.4	20.2	4.6	6.7	1.2	..
Nepal	1983/84	45.0	12.0		6.0	–	37.0[a]
	1990/91	41.2	11.8		12.3	–	34.7[a]
Republic of Korea	1985	..	49.1	..	..	..	..
	1991	..	47.4	..	..	..	..
	2001		39.7				
Thailand[b]	1985	5.7	38.1	11.9	3.2	–	41.1[c]
	1990	7.9	38.6	16.5	3.8	–	33.2[c]
	1996	11.0	38.6	18.5	4.1	–	27.8[c]
Viet Nam	1983	17.2	11.4	0.2	4.7		66.5[d]
	2010	23.1	12.8	5.1	27.0	3.8	28.2[d]

Sources: For developed countries: OECD/International Energy Agency, *Energy Policies and Programmes of IEA Countries, 1985 Revie* for developing countries: national sources.

Notes: [a] Fuelwood.

[b] Figures updated during the Committee Session.

[c] Fuelwood, charcoal, bagasse and paddy husk.

[d] Biomass.

2. Issues related to the regional energy economy pattern*
(E/ESCAP/NR.14/1)

Introduction

(i) Energy and structural change

Having surveyed sectoral final commercial energy-use patterns both cross-sectionally in a sample of developed and developing countries and longitudinally in time, looking mainly at the last 10 years of industrial energy use (tables 1 and 2), a recent report[1] conjectures that longer-term trends in energy economy patterns are characterized by an evolution from the "steady states" of least development, through developing and developed economies, to the possible future steady states of post-industrial societies. Furthermore, it is argued that during the next 100 years or so these three

types of socio-economic systems will co-exist, and the management of global interdependencies will concern itself with achieving such co-existence through peaceful means. Energy issues and structural change are viewed in a similar context in a "futures" study from New Zealand,[2] which speculates on four distinctly different futures, and considers the energy implications of social and economic change. The four "futures" could very roughly be characterized as "East", "West", "South" and "North", incorporating, respectively, the social values of (scarcity-enforced) socialism; (unbridled) capitalism; a "southern" resources (commodities) dominated, high-technology post-industrial world; and, finally, a "northern" (services) dominated, high-technology post-industrial world. What is interesting from the study is the widely diverging (though plausible) long-term energy requirements these scenarios imply for New Zealand. The household energy consumption consequences of the four scenarios are presented in the figure and table 3.

---

* Note by the ESCAP secretariat.

[1] Charuay Boonyubol and Woraphat Arthayukti, eds., *Energy Conservation in Industry – Proceedings of a Regional Training Workshop, 10-14 March 1986, Bangkok* (Energy Research and Training Centre, Chulalongkorn University, Bangkok).

[2] J.F. Boshieo and others, *Four Futures – Energy Implications of Social and Economic Change,* a report to the New Zealand Energy Research and Development Committee, Ministry of Energy, and Ministry of Works and Development, 1986.

Table 1. Energy intensities, selected countries, 1980

Energy consumption per unit of gross domestic product
(kg coal equivalent per $US 1 current GDP)

	Total	Household and commercial	Industry	Transport	Other
Developing countries					
India	.682	.314	.144	.072	.152
Indonesia	.784	.538	.104	.072	.070
Pakistan	.544	.208	.128	.076	.132
Philippines	.638	.336	.140	.068	.094
Thailand	.580	.310	.106	.108	.056
5-country average	.646	.341	.124	.079	.102
Percentage	100	53	19	12	16
Developed countries					
France	.764	.196	.214	.128	.226
Germany, Federal Rep. of	1.005	.262	.272	.144	.301
Japan	.844	.148	.322	.112	.262
United Kingdom	.744	.210	.170	.130	.264
United States	1.384	.334	.282	.336	.432
5-country average	.949	.230	.252	.170	.297
Percentage	100	24	27	18	31

Source: Charuay Boonyubol and Woraphat Arthayukti, eds., *Energy Conservation in Industry — Proceedings of a Regional Training Workshop, 10-14 March 1986, Bangkok* (Energy Research and Training Centre, Chulalongkorn University, Bangkok).

Table 2. Industrial energy consumption per unit of industrial output
(kgce/US dollar)

I. Industrialized countries (sample)

	1973	1974	1975	1976	1977	1978	1979	1980	1981	1982
France	.570	.588	.527	.531	.515	.522	.536	.534	.482	.417
Germany, Federal Rep. of	.518	.507	.462	.447	.120	.398	.410	.410	—	—
Japan	.747	.761	.761	.785	.744	.716	.698	.608	—	—
United Kingdom	.816	.787	.768	.755	.731	.728	.747	.665	.725	—
United States	1.164	1.170	1.094	1.101	1.008	.993	1.040	.990	.993	—
Average	.763	.763	.713	.724	.684	.671	.686	.641	—	—

II. Developing countries (sample)

	1973	1974	1975	1976	1977	1978	1979	1980	1981	1982
Argentina	.908	.837	.774	.626	.761	.780	.807	.836	—	—
Brazil	.726	.698	.722	.765	.833	.863	.884	.846	—	—
India	2.607	2.492	2.400	2.438	2.478	2.157	2.289	2.117	2.315	1.959
Mexico	1.289	1.265	1.070	1.119	1.031	1.149	1.211	1.079	1.100	—
Republic of Korea	1.022	1.014	1.070	1.032	1.007	1.043	1.161	1.221	1.151	.870
Average	1.310	1.261	1.207	1.205	1.222	1.198	1.270	1.220	—	—

Source: Charuay Boonyubol and Woraphat Arthayukti, eds., *Energy Conservation in Industry — Proceedings of a Regional Training Workshop, 10-14 March 1986, Bangkok* (Energy Research and Training Centre, Chulalongkorn University, Bangkok).

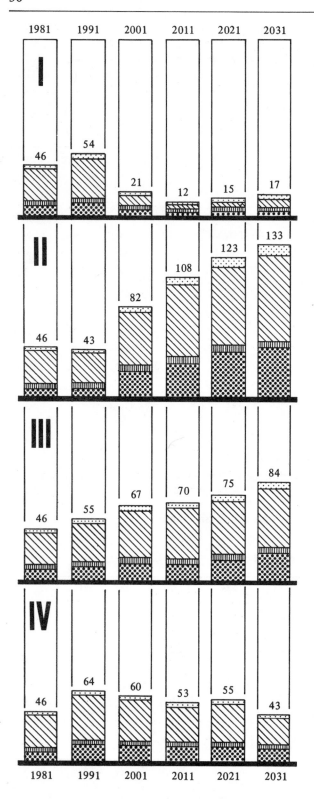

Figure. Household energy consumption (PJ/y)

Household Energy Consumption

Starting at 46 PJ/y in 1981 household energy consumption varies markedly between the scenarios by 2031. It ranges from 18 PJ/y in scenario 1 to 133 PJ/y in scenario II. These differences reflect, in part the different sizes of the populations, the relative economic performance and the different social attitudes of these two scenarios.

Table 3 presents the per capita household energy consumption for 2031. The table shows the weakness of per capita consumption as a macro-level indicator of household energy use. Economic consumption in scenario IV is over four times higher than in 1981 and yet per capita energy consumption is below the 1981 level. The crucial element which is missing in the equation when considering this scenario is the impact of technological improvement.

Table 3. Household energy consumption and economic consumption per capita in 2031 (Index 1981 = 100)

Scenario	Household energy use (1981 = 100)	Consumption per capita (1981 = 100)
I	36	82
II	292	486
III	182	165
IV	93	438

Source: J.F. Boshieo and others, *Four Futures – Energy implications of Social and Economic Change,* a report to the New Zealand Energy Research and Development Committee, Ministry of Energy, and Ministry of Works and Development, 1986.

Both scenarios (I) and (IV) could be characterized as "steady states", achieved in the first case by a drastic reduction of expectations, and in the second case through technological progress and population stabilization policies.

(ii) ESCAP work on energy and structural change

Since the thirty-seventh session of the United Nations Economic and Social Commission for Asia and the Pacific (ESCAP), the regional implications of structural changes have been studied, first in the short term on the demand side,[3] and subsequently in the light of the call by the United Nations Conference on New and Renewable Sources of Energy for a sustainable energy future through new and renewable sources of energy, on the supply side[4], [5] involving both conventional and new and renewable sources of supply. Longer-term demand-side effects were investigated in terms of energy pricing,[6] the whole process culminating in the energy and economic structural change studies presented to the first Asian Forum on Energy Policy held at Bangkok from 27 to 30 October 1986,[7] discussed below.

In the present paper regional "stock-taking" and charting of longer-term developments is presented and, then, in the last section, the recommendations of the Forum are summarized. The Committee may consider these for inclusion in the ESCAP medium-term plan for the period 1990-1995 in the light of felt regional needs and contemplated national programmes. The following paragraphs are largely taken from the report of the Forum.[8]

(a) Policy studies

(i) The four major policy studies

A team of researchers at the Asian Institute of Technology (AIT) completed the AIT study with financial support, provided under the regional energy development programme (REDP) from the United Nations Development Programme (UNDP) and the Commission of the European Communities. The Asian Development Bank (ADB) commissioned the ADB study and preliminary

results were presented to the Forum. A team of researchers at the Institute of Nuclear Energy Technology (INET) in China, the Tata Energy Research Institute (TERI) in India, and the Lawrence Berkeley Laboratories (LBL) in the United States of America completed the INET/TERI/LBL study, which was specifically commissioned for the Forum under REDP (like the AIT study). An energy pricing policy study was also commissioned under REDP and presented to the Forum by the International Labour Organisation (ILO).

(ii) Asian Institute of Technology study

The main conclusions of the AIT study were as follows:

(a) Economic structural change at a gross level almost universally tended to contribute to higher specific energy intensities;

(b) That was offset, but not fully compensated, by declining sectoral energy intensities in some middle-income countries. In low-income countries, increases in sectoral intensities accentuated the effects of structural changes;

(c) The macro-economic imbalances introduced by past oil price rises had been a principal obstacle to development, especially in low-income countries;

(d) Internal price regimes for energy had not evolved at the same rate as crude oil prices. There was evidence that price did indeed have measurable effects on energy intensities, but it was not effective in preventing the creation and continuation of imbalances;

(e) To reduce and avoid future recurrence of imbalances, energy policies would need to incorporate a larger component of demand management than had been possible in the past;

(f) Those policies would require detailed data on physical output, appliance and equipment stocks and end-use efficiency. Those data did not yet exist. Some of the proposed activities in the second cycle of REDP supported work of that type.

In the discussion that followed at the Forum, the following points were raised:

(a) Would more disaggregated data lead to similar conclusions? The lack of time series at a more disaggregate level was emphasized, as well as the necessity in certain cases to use physical, rather than monetary measures of output, since "value added" at a more disaggregated level tended to become unstable;

(b) Had time-lags of adjustments been sufficiently studied? It was emphasized that a 6-to-10 year time lag of structural adjustment to price changes had been found in a parallel AIT study that would be considered for incorporation in the final draft;

---

[3] ESCAP, *Short-term Economic Policy Aspects of Energy Situation in ESCAP Region* (Bangkok, 1981).

[4] *Proceedings of the Committee on Natural Resources, Eighth Session and of the Regional Expert Group Meeting on the Follow-up of the Nairobi Programme of Action on New and Renewable Sources of Energy,* Energy Resources Development Series No. 25, 1982 (United Nations publication, Sales No. E.83.II.F.8).

[5] ESCAP, *Energy Resources Development Problems in the ESCAP Region,* Energy Resources Development Series No. 28 (United Nations publication, Sales No. E. 86.II.F.3).

[6] C.M. Siddayao ed., *Criteria for Energy Pricing Policy* (London, Graham and Trotman, 1985).

[7] ESCAP, Report of the Asian Forum on Energy Policy, Bangkok 27-30 October 1986 (NR/AFEP/5).

[8] *Ibid.*

(c) Had commercial/non-commercial energy distinctions been taken into account? It was emphasized that sectoral commercial energy intensities had been studied, with non-commercial energy treated as an exogenous explanatory variable;

(d) Had the supply-side macro effects been sufficiently studied? The demand-side orientation had been emphasized, with supply side effects studied in terms of balance of payments and indebtedness effects of investments and energy supplies;

(e) The study was across countries, not across policies. Would a comparison of the effect of the same policy in different countries have been more instructive? It was emphasized that the one objective of the Forum was to achieve exactly such a refinement.

(iii) Asian Development Bank study

In introducing the ADB study, the secretariat emphasized that it had in fact succeeded in studying the effect of policies across countries. Common conclusions, differences and future implications were introduced in a comparative framework with the AIT study:

(a) The AIT conclusion on commercial energy intensity increasing with structural changes in developing countries was supported by the ADB study;

(b) The clear distinction between the effects of the oil price rises in low-income and middle-income developing countries, and the more price-responsive behaviour of the middle-income developing countries was also found to be the case in the ADB study;

(c) While AIT concluded that non-price demand management efforts were called for (since price measures in many cases were not sufficient), ADB still considered price measures the mainstay of adjustment;

(d) In terms of the future, since only a small portion of current energy equipment would be still around at the beginning of the next century, ADB concluded that policy should concentrate on information related to future energy equipment.

(iv) International Labour Organisation study

ILO/ARTEP (Asian Regional Team for Employment Promotion of ILO), the implementing agency for REDP activity A-1 "Energy pricing strategy", presented the main findings of the country studies organized in India, Nepal, the Philippines, Sri Lanka and Thailand. The studies varied in scope and coverage but were all broadly directed towards examining energy-pricing strategies to meet multiple criteria, including economic efficiency, meeting the basic energy needs of the poor, promoting employment, containing inflation and encouraging conservation and desired interfuel substitution. After the general framework to analyse those important issues had been developed and the research

methodologies used outlined, the main results of the country studies and their policy implications were presented. One important result concerned the impact of raising fuel prices to reflect cost factors on the economy-wide rate of inflation. Two of the studies (Sri Lanka and Thailand) concluded, after extensive analysis, that the magnitude of that effect was small. It further appeared that that might not worsen the income distribution in any marked manner. However, it seemed likely that for a number of countries the adverse effect on the low-income households could be substantial. Among commercial fuels, that effect would arise mainly through an increased outlay for kerosene and in a relatively smaller number of cases through higher expenditure for electricity. To the extent that it was possible to exercise price discrimination according to the ability to pay, the study findings suggested that there was a strong case for doing so. Where administratively feasible, income support to the low-income groups, rather than price discrimination of products, such as kerosene, would serve that purpose while also promoting conservation of energy. The studies also showed, based on sample surveys of small-scale industries as well as projections for the whole economy, that the impact on employment resulting from output and substitution effects was unlikely to be significant. The studies further showed that policies relating to the pricing of traditional fuels, in particular fuelwood, must emphasize the need for large-scale social reforestation programmes in the medium and long run, and in the short run (given that prices did not provide the right signals), to allocate more resources to produce and disseminate equipment which could considerably increase the efficiency of fuel utilization. Moreover, in pricing indigenous commercial fuels, including coal and petroleum products, too little attention was being paid to the exhaustible nature of resources. That needed to be rectified by including a realistic economic rent component in computing the cost of fuel.

(v) Institute of Nuclear Technology/Tata Energy Research Institute/Lawrence Berkeley Laboratory study

The synthesis study, based on the work of all three groups, was prepared by the Lawrence Berkeley Laboratory, Berkeley, California, using its own information and the information and comparative studies provided by the Institute of Nuclear Energy Technology, of Tsing Hwa University, China, and the Tata Energy Research Institute, India.

The study considered the evolution of energy demand in nine countries of Asia: Bangladesh, China, India, Indonesia, Malaysia, Pakistan, the Philippines, the Republic of Korea and Thailand. Together, those countries accounted for over 90 per cent of energy and oil consumption in Asian developing countries. Total gross domestic product (GDP), in 1980 United States dollars, for the study countries grew at an average of 5.8 per cent per year. Coal accounted for the largest share of energy demand; the share of oil was about 31 per cent in 1984. GDP growth rates were lower after 1979 than before. While some measures of the intensity of commercial energy use in the transport sector declined, associated with the higher prices of gasoline and diesel fuel, the intensity of commercial energy use was relatively stable in the industrial and the residential/commercial sectors. Combining those changes in intensity with structural changes and the growth in GDP, the number of vehicles and the growth of population, had led to a considerable growth in energy use (especially of coal, oil and natural gas) in the Asian economies. Similar growth in oil and overall energy demand had also occurred in the developing countries of Latin America and Africa, despite slower economic growth in those countries. The forecast of oil demand to the year 2010 for all developing countries indicated that, under some assumptions, demand might be of similar magnitude to that of OECD (Organisation for Economic Co-operation and Development) countries by that year. That high demand might lead to higher oil prices or create serious problems.

(vi) General comments at the forum on the four studies

In the light of the presentation of country papers and the papers of various organizations and research groups, the general comments that emerged could be grouped into two topics: (a) comparative studies; (b) policies.

a. Comparative studies

Intercountry comparisons could be of considerable potential use in drawing generally applicable lessons. But great heterogeneity was introduced in intercountry experience by differences in balance-of-payments conditions, in natural resources and in policy objectives, so much so that it became difficult to draw any common conclusions. The result was that most intercountry studies turned into collections of national studies loosely tied together by a general introduction.

The AIT comparative papers formed an exception to that general experience in a number of respects: they relied on more or less comparable international statistics, concentrated on energy consumption (and therefore avoided to some extent the problems raised by differences in natural resources), and sought common patterns instead of national peculiarities. Among the numerous conclusions, the following appeared to be the most significant:

(i) Structural change acted to increase energy intensity in almost all countries. Over the 10-year period studied, many countries achieved. spectacular interfuel substitution. Apart from that, achievements through energy conservation were not very significant in most countries. As a result, total energy (and oil) consumption increased significantly in almost all countries. That supported for Asia the global point made by the Lawrence Berkeley Laboratory group that while oil consumption in members of OECD had barely changed in the last 15 years, it had steadily increased in developing countries, and that if those trends continued, developing countries might outstrip industrial countries as buyers on the world oil market by the early years of the twenty-first century, and they might become responsible for increasing oil consumption. However, developing countries, including those in Asia, were lagging in energy conservation, and the more they increased energy consumption, the greater their potential for energy conservation would be. How to exploit it might thus become an increasingly important policy problem.

(ii) Energy use in transport, the most oil-intensive sector, had been strongly affected by three factors: an improvement in (or, in isolated instances, a deterioration of) the average quality of roads, a change in vehicle composition, and a rise in energy efficiency with the introduction of new models of vehicles. In going behind aggregate consumption statistics to the underlying factors, the AIT study broke new ground. The results were inferential and indicative, but nevertheless important for suggesting new directions of work.

(iii) The impression emerged, based on the vast amount of calculations embodied in the AIT papers, that the scope for further work with aggregates in value terms (e.g. sectoral gross national product (GNP)) was limited, and that inferences obtained from such data could only be tentative. Value data could not distinguish between physical and price-related phenomena. The dangers of confusing the two were suggested in the AIT studies. For instance, the relative energy intensities of different countries turned out to be very different when their GNP was compared at current exchange rates and at purchasing power parity; again, the intercountry differences in the energy intensity of the transport sector seemed to be roughly related to the relative prices of gasoline and diesel in different countries. Thus it was desirable to go behind value figures and compare physical energy intensities at the sectoral level.

(iv) If that view was correct, the scope for working with international statistics was also limited, and national data sources would have to be tapped. The 1981 ADB project had recognized that and started by making a comprehensive collection of national publications. However, in the end they were not used to make detailed intercountry comparisons.

(v) In the future, an intermediate model between those of ADB and AIT might be worth trying out. Greater resources and time could be put into national studies, closely co-ordinated and focused on exploiting and generating national data. Instead of an international team and various national teams working independently, greater interchange among them might be ensured in terms of time spent together. As such a strategy was likely to be more expensive even than the ADB strategy, it might be advisable to concentrate on particular sectors and energy uses, e.g. transport, irrigation, and power systems.

b. Policies

Policy-oriented studies might be classified into three classes, which showed international differences. First, there were official energy policy studies of various countries. They tended to be comprehensive, but because they took in the entire picture and accommodated all interests, they also tended to be relatively unfocused. Second, there were studies which had been done privately, such as those presented at the Forum. They generally relied on official papers and statistics, but often had a clearer (or narrower) view of the major energy-related problems of the countries they dealt with. Third, there were multicountry studies such as those of ADB and AIT. Those relied on a combination of international and national data sources and summarized major magnitudes in a standardized fashion. They tended to start with elementary observations on population, area, standard of living, etc. They had at least the potential of taking an independent or "objective" view of national policies, though often such a view was muffled by international inhibitions against criticism of national Governments.

Since it was Governments that made policy, the utility of the last two types of studies depended on their contribution to national policy-making. At the same time, such studies added nothing unless they involved different work or took a different view from national official agencies. Herein lay the dilemma of policy research outside Governments: it must diverge from the official viewpoint, but not so much as to lose the chance of influencing policy.

Policy alternatives could be classified into large strategic choices and detailed policy options. The strategic choices in any country were generally few and work on them was of great potential value. For instance, a country might choose in the long run between continuing to increase its already large coal production, with consequent strains on investment, technology and infrastructure, or generating nuclear power. Payments-constrained countries of South Asia had to choose between achieving a considerable increase in exports, or if exports did not increase, severely restricting transport development.[9] Resolution of such dilemmas was obviously very important, but those issues were generally very controversial, and international organizations went into them at their peril. Hence it made sense for them to concentrate on a smaller and relatively specialized area of policy. Pricing was an excellent candidate from that point of view, and its choice by both ESCAP and ILO was basically sound. A criticism of earlier published work was that it could be more operational and go into greater depth. After making all the right noises about long-range marginal cost pricing and the undesirable effects of commodity-based subsidies, the earlier studies had shied away from making clear policy recommendations because of the distributional effects of price reforms. The point was that

prices should be restructured and non-price solutions should at the same time be applied to distributional problems. The studies carried out under the REDP programme by ILO had addressed that situation by coming up with policy recommendations through their country studies. However, more work in that area needed to be done.

There were also other areas of detailed energy policy, such as investment planning, capacity utilization and demand management which could form equally fruitful areas of research.

In brief, it seemed that the broad lines of ESCAP strategy were well chosen, and that the possibilities of drastic changes in them were remote. But they could be strengthened in two ways. First, closer and more continuous co-operation between national and international research could be attempted. Second, policy studies could be made more concrete and operational, and extended to new areas.

In addition, in the energy statistics field it could be concluded that progress in the development and harmonization of energy information among ESCAP countries should continue. Solutions to the various problems which countries still faced in the collection, compilation and dissemination of energy statistics should reflect the country-specific conditions which applied. Those included.

(a) The statistical infrastructure of the country, including the role and functions of the central statistics organization;

(b) The relationship between energy ministries and energy industries;

(c) The priorities which policy makers placed on resolving particular statistical issues;

(d) Current energy data dissemination policy;

(e) The availability of trained staff.

(b) *Summary of issues and recommendations*

Issues for the medium term

The overviews of energy demand and economic structural change presented above amply illustrate the fact that while there may be certain patterns discerned in energy use and economic development, the more rational use of energy in productive activities could be one of the key factors in productivity increases. These questions have to be further investigated if successful management of economic development is our objective. The central importance of energy conservation (via organizing more efficient ways of using energy, i.e. obtaining similar or increased amounts of outputs with lower inputs of energy) lies in the fact that it might be the only way left open for labour productivity increases during the next 100 years or so, while both capital and energy might become scarce

---

[9] E.g. only if increased exports generate sufficient revenues can transport and other infrastructure-related projects materialize.

and the only recourse left is to organize ourselves better. It must be noted, however, that there are two forces at work:

(a) As countries switch from traditional to commercial energy forms — i.e. from the less efficient to the more efficient — their total energy intensity declines, but their commercial energy intensity increases;

(b) If countries develop through the expansion of (energy-intensive) industries, their energy intensity measured as consumption per capita will rise but their intensity measured as consumption per unit of output, or per unit of GDP, may go up or down. One cannot say whether a rise (or fall) in intensity is a good or a bad thing — it will depend on many other factors unique to each country. Even conservation may be a bad thing on cost grounds or because it necessitates increased imports.

Recognizing this, the Forum formulated nine recommendations, three for member country Governments and another six for concerned institutions:

(1) Based on the policy studies presented at the Forum, Governments should carefully consider the structural changes that have taken place and evolve the most appropriate measures for energy management.

(2) The management of energy data and the production of energy statistics should be organized to support the above measures and strengthened where necessary. Countries not already doing so may consider the periodic assembly of relevant energy information in a single authoritative document for the general use of Governments.

(3) The importance of traditional resources should be fully recognized and Governments should seriously address the development, distribution and better and efficient utilization of those resources, in concert with efforts to increase the income of rural people.

(4) Work on present and future sectoral energy demand should, with the co-operation of Governments, be expanded as resources permit. Work on non-commercial energy (traditional) is required in the light of scenarios of technology developments and expected economic structural changes.

(5) Further studies are needed on how to enhance the availability of non-commercial energy supplies.

(6) Further studies are needed to identify policies to make economic adjustments to future oil prices, including the current low oil price.

(7) A follow-up forum could usefully be organized to review results of the above long-term studies (and policy developments after 1986) for consideration under the second cycle (1987-1991) of REDP.

(8) In the above studies, the interests of the small and least developed countries should be duly taken into account.

(9) Direct advice on the statistical problems faced by energy administrators in different countries, including improving the consistency of information throughout the region and over time, should continue to be provided during the second cycle of REDP (1987-1991).

3. Effect of price and non-price policies on energy demand management*
(E/ESCAP/NR.14/2)

(a)   The scope of energy demand management

At the eighth session of the Committee on Natural Resources held in October 1981, the paper "Energy transition through demand management: long-term strategies and policy imperatives" was presented by Professor Ramesh Bhatia of the Institute of Economic Growth, New Delhi.[1] Professor Bhatia outlined the need and scope for demand management and proceeded to discuss strategies of demand management, including strategies to influence final demand, strategies to influence the input-mix of an economy, pricing policies and non-price instruments, taxation, incentives, information and management, as well as technology.

In subsequent years, many countries have instituted various demand management-related policy measures. These have been reviewed by a recent Asian Development Bank (ADB) study,[2] giving an overview of lessons learned in a sample of countries.

An outline of the sectoral demand analysis and rational use of energy programme of ESCAP is presented (see annex II) as an attempt to draw programmatic conclusions from the lessons learned in a co-ordinated intercountry programme of supporting activities to help countries of the region achieve a sustainable energy future.

---

*   Note by the ESCAP secretariat.

[1]   *Proceedings of the Committee on Natural Resources, Eighth Session and of the Regional Expert Group Meeting on the Follow-up of the Nairobi Programme of Action on New and Renewable Sources of Energy,* Energy Resources Development Series No. 25, 1982 (United Nations publication, Sales No. E.83.II.F.8), p. 55.

[2]   "Review of energy demand management in nine oil-importing developing member countries of the Asian Development Bank", paper submitted to the first Asian Forum on Energy Policy, Bangkok, 27-30 October 1986.

(b) Energy demand management measures and impacts in a sample of countries of the region

During the period 1984-1986, the Energy Planning Unit of ADB undertook a review of energy demand management in nine oil-importing developing member countries of the Bank. The results were presented at the first Asian Forum on Energy Policy (Bangkok, 27-30 October 1986). The lessons learned from these results have been utilized in current and planned follow-up action by the ESCAP secretariat, as described in section (c) below.

On the policy level, two related issues should be noted, in particular in growing economies. First, the production capacity of today will, owing to growth and expansion, account only for a minor share after 10-15 years. In other words, the major part of the production capacity at the year 2000 has not yet been installed. Efforts have to be made to ensure that new capacity incorporates modern, energy-efficient equipment and processes. Second, changes in the structure of the industry can have a major impact on the energy demand of the industrial sector. In the long term, these developments will have a far greater impact on industrial energy demand than efforts to improve the efficiency of existing plants.

(c) Summary of conclusions and recommendations: a proposed co-ordinated intercountry programme in energy management

Having reviewed the energy demand management and conservation policy achievements in a sample of countries of the region and in the light of the recommendations of the first Asian Forum on Energy Policy, the ESCAP secretariat initiated a three-year programme to achieve a regional co-ordinated approach to the rational use of energy with a longer-term time horizon of about 15 years, allowing sufficient time for national programmes to achieve carefully-formulated sectoral targets.

Assistance in achieving this is given by the regional energy adviser (see annex I) whose work has already been instrumental in drawing up sectoral demand management plans in some countries, as well as the regional energy statistics adviser, whose help in identifying information gaps and formulating programmes for overcoming these may prove valuable for developing countries of the region. The services of these advisers are available upon request by member Governments.

In addition, under the regional energy development programme (REDP) for Asia, with the support of UNDP, the German Agency for Technical Co-operation (GTZ), and the Commission of the European Communities, a three-year implementation plan has been drawn up for assistance in strengthening national capabilities for energy planning and data management (see annex II).

In reviewing this implementation plan, countries of the region are asked to indicate their interest in participating and in contributing to the achievement of its objectives. It must be borne in mind that a certain commitment of national resources to the programme is required to render it operational. These requirements have been estimated for the initial sample participating countries and were discussed at the recent tripartite review of the regional energy development programme. The Committee is now requested to endorse, in principle, the recommendations made in the implementation plan of this project; the actual phasing and budgeting of in-country projects are to be decided by in-country project teams.

Annex I

ESCAP ADVISORY SERVICES ON ENERGY

Advisory services on energy rendered by visits by the regional adviser on energy based at ESCAP, Bangkok are available free of charge to developing members of ESCAP upon request from the Governments through their respective UNDP offices.

The services in the energy field include the formulation of energy development programmes, energy management and conservation (rationalization of use), audits of industries and power systems, testing of power machinery and industrial installations, programming and development of research and technical training, etc.

Requesting Governments should specify the scope of assistance required, duration (should be no more than six weeks) and proposed timing for the mission, and submit the information to the respective UNDP office.

Annex II

"COMPREHENSIVE IMPLEMENTATION PLAN" FOR REDP ACTIVITY P-1.1

"Assistance in strengthening national capabilities for energy planning and data management — sectoral energy demand studies"

I. Background

As recommended by the self-evaluation study of sub-programme 1, "Strengthening national capabilities in integrated energy planning and programming", REDP second cycle activity P-1.1, "Sectoral energy demand studies" is considered essential for achieving the objectives of the subprogramme as laid down in the medium-term plan: "To assist regional developing countries by strengthening their national capability in preparing and carrying out comprehensive energy development programmes" leading to "regionally co-ordinated national energy plans in the context of overall economic development plans and with adequate consideration to demand management". Previously, when a similar activity was scheduled in 1984-1985, the project could not be carried for lack of extrabudgetary funds. With UNDP (REDP) funding, however, preparatory activities leading to the October 1986 first Asian Forum on Energy Policy were successfully completed, re-emphasizing the need for more work in this area. Accordingly, the new 1987-1991 REDP cycle II project document contains a comprehensive multi-sectoral intercountry energy demand analysis activity with linked studies and training components, in a comparative context, that is likely to achieve the objectives of the subprogramme by the end of 1989. Detailed below is the first draft of a plan for implementing this activity, for approval by the Tripartite Review Meeting of the Regional Energy Development Programme in August 1987, and for subsequent in-country resource allocations until the fourteenth session of the Committee on Natural Resources at the end of October 1987, based on the recommendation of the Tripartite Review Meeting.

II. Main features of the implementation plan

1. It is expected that the results of previous studies/experience will be fully utilized;

2. It is expected that the orientation will be towards sectoral energy demand management to satisfy future needs (10-15 year time horizon);

3. It is expected that the activity will achieve medium-term plan objectives by either harmonizing energy plans by 1989, or setting up a mechanism for continuous adjustment of national energy plans in a harmonized manner through regular consultations by national energy focal points.

III. Activity milestones and work programme

1. Preliminary choice of countries by ESCAP/GTZ team — February 1987

2. In-country consultants hired by ESCAP and GTZ for chosen countries to formulate "indicative 15-year sectoral demand management alternatives" — March 1987

3. Lead consultant and ESCAP team visits selected study countries to:

   (a) Discuss progress

   (b) Discuss formation of in-country teams and their training needs for possible follow-up detailed sector-specific (or area-specific) studies to fill data gaps

   (c) Collect information for synthesis report — May 1987

4. Lead consultant writes synthesis report based on seven (five ESCAP, two GTZ) in-country reports plus possible additional reports by the regional energy statistics adviser — June-July 1987

5. Synthesis report (overview paper and comprehensive plan for follow-up studies) completed by main consultant and submitted to:

   (a) Natural Resources Division (for the fourteenth session of the Committee on Natural Resources)

   (b) REDP office for distribution to focal points in preparation for the Tripartite Review Meeting.

June-July-August 1987

6. Tripartite Review Meeting (26-28 August 1987) discusses synthesis report and identifies necessary in-country detailed study budget requirements and training needs for critical sectors, based on recommendations of synthesis report. Meeting makes budgetary recommendations for participating country consideration. Report of Tripartite Review with recommendations transmitted to participating country seats of government by mid-September 1987 (by ESCAP)

7. The report of the Tripartite Review Meeting, as well as the synthesis report, presented to the Committee on Natural Resources for decisions on participation/budgetary allocations by member countries — 27 October-2 November 1987.

8. Synthesis report submitted for finalization (and advice on technical editing) at the 5-9 October 1987 Expert Group Meeting that will:

    (a) Finalize the ESCAP/GTZ study.

    (b) Agree on detailed in-country follow-up (contingent on budget availabilities) for 1988.

9. ESCAP/GTZ publish synthesis report and indicative plans as joint volume under the title "Policies for rational use of energy".

10. In-country teams formed and formal contracts signed – November-December 1987.

11. In-country teams perform detailed country-specific analysis of critical sectors as identified by ESCAP/GTZ study (ESCAP/GTZ advisory team available on request to assist in-country teams).

12. Progress meeting (expert TCDC working group) called to discuss results and compare experience – September 1988.

13. Second tripartite review evaluates progress and guides in-country follow-up – September 1988 (organized in conjunction with expert TCDC working group).

14. In-country teams prepare final reports for action-oriented national workshops, based on:

    (a) Indicative 15-year plans.

    (b) Sectoral in-country study findings.

October 1988-December 1989

15. National workshops organized – February-March 1989.

16. Main consultant/ESCAP prepares final evaluation report and recommendations for intercountry co-operative follow-up – February/March/April 1989.

17. ESCAP organizes "Forum on energy issues in development planning" based on final evaluation report in conjunction with 1989 Tripartite Review – September 1989.

18. Technical editor finalizes entire project consolidated report for publication – October-December 1989.

IV. Issues of co-ordination

1. In all the above it is assumed that:

(a) The ESCAP/GTZ project now under way will be specifically used to prepare the ground for this REDP project, at no extra cost to UNDP, thus achieving a certain degree of co-financing (similar co-financing with the Commission of the European Communities could, perhaps, be negotiated from 1988 and the lead consultant thus changed for 1988 from GTZ to CEC, if mutually agreeable?)

(b) The following ESCAP Energy Resources Section work programme elements will be devoted to the accomplishment of the objectives of REDP activity P-1.1, at the same time satisfying all the criteria for the success of the objective of subprogramme 1 of the medium-term plan, 1984-1989, which is the strengthening of national capabilities in integrated energy planning and programming.

1987

1.3 (ii) "Report to the Committee on Natural Resources on effects of price and non-price policies on energy demand management".

1.3 (iv) (c) "Advisory missions on policies and strategies in the field of energy demand management".

3.3 (i) "Study on energy conservation policy and measures for energy demand management".

3.3 (ii) "Workshop on energy conservation policy and measures for energy demand management".

1988-1989

1.2 (iii) (b) "Project on strengthening national capabilities in energy planning in a rapidly changing economic environment".

2. The regional energy adviser, the regional energy statistics adviser plus staff of other divisions concerned with specific sectoral studies (industry, transport, agriculture and domestic/commercial) will be available to assist in the project as and when required.

4. Bangladesh: energy situation, demand management and energy policy
(Synopsis)

(a) Energy situation

Traditional energy has overriding importance in meeting the energy needs of almost 65 per cent of the total population that live in the rural areas of Bangladesh. Use of commercial energy comprising only about 35 per cent of the total energy (17 per cent by natural gas and 15 per cent by imported oil) has bright economic prospects. Of the total 11.3 million tons of oil equivalent (TOE) of energy use in the country, primary commercial energy provides about 4.3 million TOE. 1.6 million tons of crude oil and petroleum products were imported in 1986, which took away about 35 per cent of the total export earnings of the country. On the other hand, through efficient use of indigenous natural gas, the country saved about 2.4 million TOE in the same year. Consequently, the demand management of commercial energy essentially rests on effective and efficient natural gas utilization.

It is of interest to note that the country is divided geographically into two parts by a major river called Jamuna. Because of this physical separation, practically two separate energy demand management scenarios have developed in the country. In the short term, the Government of Bangladesh has therefore placed adequate emphasis on the transmission and distribution of gas in the Eastern Zone. In the Western Zone, on the other hand, fuel for generation of electricity and other commercial and industrial use comes from imported oil. In the mid to long term, with regard to the overall energy prospects of the country, other available options and potentials, like exploitation and marketing of coal and peat discovered in the Western Zone, are being seriously examined. The demand forecast also shows that gas use in the commercial energy sub-sector will rise from the present contribution of 58 per cent in 1987 to about 70 per cent by 1990.

At present 38 per cent of natural gas is being consumed for generation of electricity, 11 per cent as fuel in industries, 35 per cent as feedstock and fuel in fertilizer production, 7 per cent for brick burning and 9 per cent as fuel in domestic and commercial areas. This trend may continue up to 1990, except for increased use in electricity generation. While the use of imported petroleum products heavily rests on the transport sector (38 per cent), industry (22 per cent), and commercial use (24 per cent), only about 10 per cent is used in generation of electricity and 6 per cent for household purposes. It is expected that through careful demand management, under the overall primary commercial energy scene, the contribution of natural gas will rise substantially and the use of petroleum products will remain more or less stable. The incremental primary commercial energy demand would be met from gas in the short run and coal, peat and others in the medium to long run.

Regarding new and renewable energy development a three phase action programme has been envisaged. This action-oriented design is not intended to undertake basic research which duplicates work done elsewhere or to generate general studies on the 'state of the art' of various technologies, or the like. The near-term objective would be to develop local capabilities, install field testing and demonstration units, assess market demand and economic viability, and where appropriate (such as with improved and more energy-efficient "chulas") undertake some initial distribution. Another output of the project would be a work programme identifying the most promising technologies for large-scale dissemination, the detailed implementation steps and investment requirements. The work plan can constitute a follow-up project suitable for international funding.

(b) Energy demand management

The first oil crisis in November 1973, and to an even greater extent the second oil price hike of 1979/80, confronted the non-oil producing Bangladesh with a new and more difficult energy scenario. The problem for the Government was to face uncertainties and make *ad hoc* adjustment plans for future unexpected increases in international oil prices. Although the country consumes an insignificant amount of petroleum products, dependence on the same could not be overcome, nor could conservation of energy in general be introduced. However, substitution by other sources of energy for imported oil did take place, at least in the primary commercial energy sub-sector. Fuel switching has been achieved mainly by developing indigenous natural gas.

Since 1983 the oil prices have been showing a downward trend. Therefore, it has become necessary for the State Corporation, Bangladesh Petroleum Corporation, to address itself to some fundamental questions like whether, and to what extent, the policies adopted in response to high oil prices should be passed on to consumers, after recovering the huge losses sustained during the mid-1980s due to non-adjustment of petroleum product prices in the domestic sector, or should domestic oil prices be kept high in order to stimulate continued energy conservation. The solution to these and related questions is yet to be achieved.

Despite the low level of energy consumption in Bangladesh, the present energy supply situation to meet the growing needs of the country is less than satisfactory. The rural population is overwhelmingly dependent on the non-commercial sources. The demand for commercial energy has been growing very fast in recent years. Per capita energy, which was recorded as 24 kg of oil equivalent (kgoe) in 1972, rose to 31 kgoe in 1983 and 42 kgoe in 1986, and is expected to reach 58 kgoe in 1990. Electrical energy generation is projected to increase from 4,536 GWh in 1984-1985 to 9,262 GWh in 1989-1990. It is also estimated that the use of commercial energy will grow at 8.7

per cent per annum, while the rates of growth in the gas sector and electrical energy sector would be 12.7 and 15 per cent respectively. With the introduction of more energy-intensive agricultural practices and technologies to increase food production and expansion of the small and cottage industries in the future, the rural electrification programme will continue to enjoy priority and further expand.

It has been observed in the recently concluded Asian Development Bank study entitled "Energy policy experience of Asian countries" that over the last 10-year period (1973-1983), the share of non-commercial energy in the total primary energy supplies of Bangladesh declined significantly (11 per cent). At the same time the import dependence of Bangladesh decreased by about 12 per cent during the same 10-year period.

As regards changes in energy and oil intensities, after 1973, the total energy intensities of Bangladesh declined. However, after the second crisis, the total energy intensity of the country rose moderately; the commercial energy intensity of the country rose, but because of the increased use of natural gas, its oil intensity declined.

(c)   Impact of changes in international oil prices

In 1973, before the first oil crisis, Bangladesh had a trade deficit of $US 540 million. The crude oil import bill in that year was only $US 21 million. The trade deficit rose to $US 625 million in 1974 and $US 848 million in 1975 and dropped to $US 430 million in 1976. The cost of oil imports was higher in those years, but it did not move in step with the trade balance and accounted for only a fraction of the trade deficit.

In 1973 Bangladesh was in the grip of severe inflation. The consumer price index (CPI) rose to 45.6 per cent in 1973, went up to 54.1 per cent in 1974. The CPI declined 9.5 per cent in 1976, and from 1978 to 1981 it levelled off at an annual rate of increase of about 12 to 13 per cent. The gross domestic product (GDP) behaved erratically, increasing over 12 per cent in 1974 and 1976, but growing only 3.4 per cent in 1975 and 1.3 per cent in 1977. The GDP continued its erratic behaviour, increasing 6.5 per cent in 1978, 4.6 per cent in 1979, 1.3 per cent in 1980, and 6 per cent in 1981.

The trade deficit continued to rise from 1977 until it peaked at $US 1,644 million in 1981. The net deficit in trade with oil-exporting countries exceeded the cost of oil imports in three of the five years from 1977 to 1981.

Both international oil prices and the trade deficit of Bangladesh declined in 1983, but these developments were not closely connected. The CPI increased 9.3 per cent in 1982 and 8.1 per cent in 1983. Despite the stimulous of lower international oil prices, GDP grew only 0.8 per cent in 1982 and 3.3 per cent in 1983. The scenario is changing now.

The Government of Bangladesh regulates oil, gas and electricity prices as part of general price control for achieving other economic and social objectives. The distribution of electricity, petroleum products, and natural gas is in the hands of government monopolies. Earlier the main consideration in setting energy prices was maintaining positive cash flows for the government-owned energy companies. Price changes required government approval.

The Government of Bangladesh delayed adjustments in imported oil prices during the early 1980s, causing serious financial problems for the oil company and the electric power company. From 1972/1973 to 1982/1983, the price of crude oil increased 9.71 times, but the index of real commercial energy prices increased only 3.82 times. Domestic prices of oil and natural gas increased 4.01 times and 3.23 times respectively. The Government also kept some energy prices lower than others. The prices of gasoline and high-speed diesel oil were raised less than the prices of other petroleum products.

Natural gas for electric power and fertilizer plants was sold at special low prices. The substitution of natural gas was brought about by making gas available to more consumers, rather than by selling gas more cheaply than oil. Government has taken a view to consider the long run marginal cost (LRMC) method and is accordingly setting the gas tariff structure now.

(d)   Energy conservation measures

The Energy Monitoring Unit (EMU), which originated from a joint concern by the Government and the World Bank about inefficient energy use in Bangladesh, was established within the Ministry of Energy and Mineral Resources to develop, initiate and carry out a national industrial energy conservaton and diversification programme to improve energy use efficiency in the industry and power sector.

The main responsibilities of EMU include policy co-ordination, planning, promotion and monitoring of energy conservation to establish an efficiency improvement programme focusing on energy audits of plants in major energy-using industry sectors like fertilizer, power, pulp and paper, jute, textiles, steel etc. Over 200 energy conservation/efficiency improvement projects have been examined. The results of the audit also indicate a very significant potential for industrial energy conservation/efficiency improvements.

As a part of the regulatory activity, the most drastic measure applied was area-wide load shedding during planned or forced outages of generating units. To reduce peak loads, holidays for major industries were staggered, and the hours of operation of some industries were restricted. A high tariff was charged for the use of electricity for ceremonial purposes, and such use was temporarily prohibited. Electrical shop displays and advertising signs were also discouraged. The use of air conditioners in government offices was reduced.

A secondary objective of energy conservation regulations was to save gasoline. Import duties on motorcycles and scooters were lowered, duties on cars were graded according to engine size, and the importation of cars whose engine capacities exceeded 1,300 cc was banned. The use of government vehicles was sharply restricted, and the purchase of new vehicles required the approval of the highest authority.

In the absence of time-of-day metering, restrictions on the use of electricity during peak hours proved to be unenforceable. Efforts to stagger holidays were also widely ignored. Some weddings and other social functions continued unabated. Substantial gasoline savings were achieved, but this increased the surplus of gasoline being exported at very low prices.

(e) Policy framework

Bangladesh is a low energy using, subsistence economy, whose energy use is dominated by non-commercial biomass fuels. Economic development of the country will entail changes in this pattern of use. Increased agricultural output will require the intensification of agricultural production because of land availability constraints. Economic growth and increased industrialization will lead to an increase in the energy intensity of the economy. This will reflect itself in higher energy demands, particularly for commercial fuels, which will be accentuated by a lagging availability of biomass fuels. The bulk of the increase in demand can be expected to come from the productive, as opposed to the domestic, sectors of the economy.

Population growth has a direct bearing on energy demand in the domestic sector. Also the rate of urbanization is an important factor in determining the amount and pattern of energy use in this sector. Any improvement in living standards would have significant consequences for energy demand. Thus to convert aggregate energy demand into fuel requirements, energy policy considerations shall have to be introduced.

Since the rural energy development plans cover areas as yet not institutionalized, proper funding must be made available to support the phases of the envisaged rural and new and renewable energy programme. Increasing amounts of energy will be required in the rural areas, mainly for agricultural applications (irrigation, fertilizers, machinery and transport) but also in domestic households to cook food for the growing population. Much of the future increase in energy demand will have to be met by commercial fuels.

In addition a nuclear energy policy has emerged, even though the nuclear option does not appear economic in the period to 2000 based on current cost estimates and the Bangladesh Energy Plan's world energy price projections. Perhaps it would be prudent for Bangladesh to retain its capability (with the Atomic Energy Commission) to develop a nuclear energy option in the long-term.

The second oil crisis in 1979/80 caused the Government to tighten controls over credit, impose new taxes, reduce subsidies and other export incentives. Reactions in Bangladesh toward energy demand management were more positive following the second oil crisis than after the first crisis. The second five-year plan (1980-1985) called for the substitution of indigenous fuels for imported fuels, improvement of the electricity distribution system, expansion of the gas transmission and distribution system, increasing gas production, stepping up exploration for oil and gas, and developing renewable energy sources including community forests. The target for the supply to petroleum products was sharply reduced.

In addition to the five-year plans, energy policy has been articulated principally in the ADB/UNDP funded Bangladesh Energy Study of 1976, and the National Energy Policy, approved by the Council of Ministers in September 1980. The study was a commendable pioneering effort, but it was based on weak data and made some erroneous assumptions. Although the National Energy Policy was not based on rigorous analyses, it contained useful prescriptions for managing the energy sector and also for diverting increases in energy demand from oil to gas. The recently concluded Bangladesh Energy Plan study funded by ADB/UNDP is currently under review of the Government.

5. The production, consumption and policy of energy resources in China
(Synopsis)

In the past few years, in virtue of the open policy of adjusting, reforming and rejuvenating the national economy, the energy resources industry in China has witnessed steady growth and changes in the consumption structure, increase of cost-effectiveness in energy use, alleviation of the problems in the demand and supply of coal, and further improvement of rural energy resources. However, the short supplies of electricity and petro-chemical products are limiting factors for economic development, and the Government has been redoubling its efforts in the area. The following is some of the basic information with regard to the production, consumption and policy of energy resources.

(a) Energy production

Since 1982, energy production has been increasing continuously. During the five years from 1982 to 1986, the annual rates of growth were 5.6, 6.7, 9.3, 9.8, 3.8 per cent respectively. In 1986, energy production reached up to 888 MTCE, with an increase of 32.9 per cent or 32.2 MTCE within five years.

In the course of five years, with high growth from 1982 onwards, the annual output of coal production registered an increase of 44.69, 48.21, 74.70, 83.05 and 212.76 million tons respectively over the preceding year. The gap between the demand and supply had been narrowed, and the efficacy of coal production had been on the rise. Further mechanization of mining had greatly ameliorated safety in production. A number of new mines equipped with advanced facilities, for example, the open-air coal mine in Pingshuo, a joint-venture enterprise, had been put into operation.

Oil production entered a new stage of development after years of stagnation. The oil output began to pick up in 1983, and it had increased by 8.1 per cent in 1984, 9 per cent in 1985, 4.9 per cent in 1986 compared with the preceding year's output. The average annual output was 131 million tons. The growth of natural gas production was comparatively lower with the annual output remaining approximately at 13 billion cubic metres for years.

In the past several years, efforts have been made to finance the development of electric power plants including the promotion of small-scale hydropower. During the period 1982-1986, the production of electricity has registered an increase of 37.2 per cent, of which that of hydropower was 27 per cent, the annual rates of increase were 6, 7.2, 7.2, 8.9, 9.4 per cent respectively. The generated power in 1986 was 449.6 TWh, of which the hydropower generation was 94.5 TWh. However, despite the increase in power generation, it still fell short of demand.

(b) Energy consumption

In the wake of the development of the national economy and the improvement of the standard of living, energy consumption had correspondingly augmented. During the 1981-1985 sixth five-year plan, the total energy consumption was 3.36 GTCE, or 600 MTCE more than that of the 1976-1980 fifth five-year plan, the average annual rate of increase being 5 per cent. The major structural changes of the energy consumption due to oil conservation were as follows:

(1) While the share of coal increased in the energy consumption, that of oil declined. During the period, 1980 to 1985, coal increased from 72.1 to 75.92 per cent, oil decreased from 20.85 to 17.02 per cent, natural gas decreased from 3.06 to 2.23 per cent, hydropower increased from 4.03 to 4.83 per cent.

(2) The share of energy consumption in the manufacturing sector was reduced, whereas domestic consumption was on the increase. During 1981-1985, the average increment of energy consumption in the manufacturing sector was 4.1 per cent, and that of domestic consumption was 8.8 per cent.

(3) The energy consumption within the industrial sector has been undergoing structural changes in terms of the energy efficiency pattern. During the period 1981-

1985, China had accelerated the development of the light industry and textile industry, and reorientated the service and the structure of products.

(c) Improvement of energy efficiency and great achievements in energy conservation

In recent years, energy efficiency has been greatly ameliorated as a result of adopting the policy of integrating energy conservation with the development and restructuring of the economy and enhancing energy management and the technical innovation in energy conservation. The per capita national income had increased by 24.3 per cent from 612 yuan in 1980 to 761 yuan in 1985. In the course of five years, the conserved energy amounted to 149 MTCE. During the period of 1980-1985, the 3.9 per cent average annual increase of energy consumption in the industrial sector had fuelled a growth rate of 10.8 per cent in the industrial production. The energy consumption for every ten thousand yuan of industrial output had diminished from 7.84 MTCE in 1980 to 5.69 MTCE in 1985, with an average annual decrease of 6.2 per cent.

(d) New development in rural energy resources

From 1981 to 1985, firewood forests had been expanded by 35 million mu (15 mu = 1 ha), the installed capacity of small hydropower had been raised by 2.5 GW, and a total of 2.5 million biogas units had been constructed. Thirty-five million families had benefited from the energy-efficient firewood stoves. In line with the geographical diversities, efforts had been made to further improve the rural energy situation by developing solar, geothermal and wind powers.

(e) China's current policy of energy

(i) Coal

Continued efforts will be made to further develop large, small, and medium-sized coal mining and mobilize all the efforts at national, collective and individual levels. Both the local authorities and the sectors will be encouraged to finance coal mining. The state-administered coal mines will adopt the general contracting system for the input and output, and maximize the output with even less input. Restructuring will be undertaken in the coal mines owned by the small towns. The existing coal mines will carry out technical innovations and facilitate mechanization so as to improve the operational equipment, enhance safety in production, and increase productivity.

(ii) Oil and natural gas

It is pivotal to strengthen exploration of oil and natural gas and increase their reserves. Efforts will be made to overcome the unbalanced development of oil and natural gas. With regard to the existing oil and natural gas fields, it is essential to systematically improve the infrastructure and undertake projects of reconstruction and extension in order to maintain and raise production. Co-operation with foreign countries will be carried out to introduce and adopt advanced technology and management from abroad. The Government has formulated and implemented the policies of assigning responsibility for output quotas with a view to mobilizing the initiatives of the enterprises and the workers.

(iii) Electric power

The development of the energy industry will be focused on electric power and accelerate the establishment of thermal power plants. It is planned to construct a number of large-scale thermal power plants at the coal bases. Co-ordinated efforts will be needed by the power users to set up thermal power plants in the well-accessed offshore regions. It is requisite to develop hydropower, particularly along the segments of a river, where the potential hydropower is rich, location is ideal and the possible loss caused by the installation is minimum. Nuclear energy will be explored gradually. Small hydropower and thermal power with geographical advantages will be developed. Reforms will be undertaken in the financing systems. It is necessary to stimulate the financing for the development of power generation at the local, sectorial and enterprise levels, and break away from the stereotyped practice, by which only the Government invested in and managed the power generation business.

6. The energy development in Indonesia
(Synopsis)

(a) Energy policy

The National Energy Policy is derived from the Main Guidelines of State Policies (GBHN) as well as from the development policies of the five-year development plan (REPELITA).

The major objectives of the energy policy are:

(a) To assure a gradual and smooth transition from a mono-energy (namely oil) to a poly-energy economy;

(b) To assure the availability of energy for the domestic market at reasonable prices;

(c) To ensure a continous and positive contribution to the balances of payment and public revenue.

Accordingly, the following measures will be taken:

(a) Intensification, for example, to increase and expand surveys and explorations of energy resources, conventional (oil and gas included), as well as non-conventional;

(b) Diversification through reduction of dependence on oil in the overall energy consumption, by developing and using non-oil energy resources;

(c) Energy conservation by economizing energy use and using energy efficiently and wisely;

(d) Indexation, by matching each energy need with the most appropriate energy source available in the country.

(i) Intensification

Exploration efforts for energy resources has been most affected by the decline of oil prices. Nevertheless activities in the past years have resulted in new discoveries which could provide several options for meeting the energy requirements of the country.

Petroleum. There are 60 sedimentary basins, both onshore and offshore. Of these, 18 basins have been explored fairly and more than one half are virtually untouched. Ten basins are now producing oil and/or gas. The success ratio, in the exploration campaign remains satisfactory and is about 30-40 per cent. It is estimated that of the total undiscovered resources of hydrocarbon at least 50 billion barrels of oil equivalent (BOE) recoverable of crude oil remain to be discovered. The proven recoverable oil reserves have been estimated at about 20.1 billion barrels (bb) of which about 507.2 million barrels (mmb) (including condensate) have been produced by 1986. Current production is about 1.3 mmb/d. Exploration and development expenditures are at an all-time high and are still growing. Secondary recovery and enhanced oil recovery are under way; the political, social and investment climate continues to be stable; therefore capacity to produce oil can grow to 1.8-2.0 mmb/d in the future.

Natural gas. A recent estimate prepared by the Indonesian Geological Society in 1985 accounts for 217 Trillion cubic feet (Tcf) of recoverable natural gas. The estimated remaining proven recoverable reserves is 80 Tcf, mostly located in the South China Sea (the Natuna fields), Kalimantan and North Sumatra, of which 85 per cent are non-associated and can be produced independently of oil. Production of natural gas reached 1.58 Tcf in 1985 and 1.63 Tcf in 1986; 56.2 per cent of this is produced to fulfil export agreements of liquified natural gas (LNG).

Coal exploration programmes carried out by the Government in South and Western Central Sumatra and by private foreign contractors in East and South-eastern Kalimantan, have for the greater part been completed. Proven coal reserves to date amount to some 1.5 billion tonnes, while total indicated and inferred reserves amount to another 6.25 billion tonnes. Besides, geological resources of coal are estimated at some 15.9 billion tonnes. Indonesia's coal production increased from 1.467 million tonnes in 1984 to 1.943 million tonnes in 1985 and around 2.940 million tonnes in 1986, while its exports of higher grade coal and anthracite amounted to 1.033 million tonnes in 1985 and some 1.3 million tonnes in 1986.

The *hydropower* potential of Indonesia has been estimated at 78,000 MW or 410,000 GWh/year. Some

34,000 MW would be developed to build large-scale hydro-electric plants, of which about 2,153 MW had been installed by late 1986. Small-scale hydropower could be utilized to generate electricity for rural and remote areas. Based on a recent study, in West Java alone about 20 GJ/year of micro hydroelectricity could be developed.

Exploration of *geothermal* potentials has been continued. The total exploitable potential has been estimated at 10,000 MW, with about half of that on Java. Some of these which have been investigated have a potential of approximately 3,300 MW. At present, there is one 30-megawatt geothermal unit in operation at Kamojang, West Java and one 2-megawatt unit on the Dieng Plateau of Central Java.

Biomas includes forest biomass (113 million ha, which is 59 per cent of total land area) and agricultural waste. In 1984 the production of wood including firewood was 165 million m^3. Forest waste was about 30 to 40 per cent, or about 15 million m^3/year, 50 per cent of which was from Kalimantan. Estimated density was 354 kg/m^3 and its heat content was 3,500 kcal/kg. With a 25 per cent efficiency these could produce 27,000 GWh of electricity. Other wood waste was about of 30 to 35 per cent of forest and 3 million m^3 from plywood factories. Agricultural waste includes crop waste (from rice, maize, cassava, sweet potatoes, peanuts, soybean, coconut, coffee); waste from plantations (mill effluent:palm oil); and animal residues. A recent assessment on the potential of agricultural waste indicated an annual amount of 29.5 million tonnes. From 23 million tonnes of rice estimated to have been harvested in 1983, an estimated 4.7 tonnes of chaff could have been collected as feedstock for energy.

A great number of livestock such as buffaloes, cows, and others have been assessed to obtain adequate volume of dung for biogas production. Annual dung production amounts to about 114 million tonnes.

There has been intensive and continuous efforts for a regeneration of degraded land areas and for firewood supply through a reforestation program.

Peat is available in Sumatra and Kalimantan with resources of about 200 billion tonnes. This was estimated from field observations over areas covering a total of approximately 17 million hectares. Studies have been started to utilize peat in the near future as a substitute for firewood for domestic use and for medium to large-size power generation. In Indonesia pilot projects on peat utilization will be introduced.

There are large variations of *solar energy* intensity over the country. From various surveys, it is known that the average annual solar radiation in Indonesia is between 1,668 kWh/m^2 and 1,946 kWh/m^2. The average effective radiation is approximately 6 hours/day.

As with solar energy, *wind power* is still used in the traditional way such as in the traditional mode of coastal fishing and inter-island shipping. According to available data, several locations have average recorded wind speeds of 20 km/hr or more with an intensity of more than 1500 kWh/m^2/hr.

Being an archipelago, Indonesia has the potential for ocean thermal energy conversian (OTEC). Several areas such as Nauru and Bali have shown that OTEC can be used to generate electric power.

(ii) Diversification

The development of non-oil energy resources has been accelerated during the past several years, which resulted in the increase in the use of non-oil energy sources such as natural gas, coal, geothermal and hydropower. Table 1 shows the commercial energy supply mix during the first three years of the REPELITA IV (1984/85-1988/89). It can be seen from the table that the share of non-oil energy sources increased from 27.10 per cent in 1984/85 to 34.12 per cent in 1986/87 while the share of oil is becoming smaller and smaller. It is projected that by the end of REPELITA IV the share of oil will decrease to 62.43 per cent.

Table 1. Commercial energy supply mix
(1,000 barrels of oil equivalent)

Energy source	1984/1985				1985/1986		1986/1987		1988/1989	
	Projection	Percentage	Realization	Percentage	Realization	Percentage	Realization	Percentage	Projection	Percentage
Coal	5 201	2.34	1 967	1.81	6 913	3.58	20 372	7.52	28 244	6.66
Hydropower	16 094	7.26	14 017	6.15	17 662	7.53	21 090	8.56	24 330	8.32
Geothermal	515	0.23	433	0.19	448	0.19	464	0.19	1 958	0.67
Natural-Gas	40 805	18.41	45 314	19.89	47 434	20.21	53 688	21.79	25 246	18.91
Sub-total non-oil	62 615	28.25	61 731	27.10	72 457	30.88	84 063	34.12	109 778	37.57
Oil	159 032	71.75	166 039	72.90	162 195	62.12	162 319	65.88	182 408	62.43
Total	221 647	100	277 770	100	234 652	100	246 382	100	292 186	100

(iii) Energy Conservation

The Government, through energy conservation programmes, intends to increase the efficiency of energy use in all energy consuming sectors. Industry is one of the sectors which offer a great opportunity and a substantial saving potential. Not less than 112 industrial plants have been audited for their energy use for the last 3 years. The result of this energy audit has indicated that the potential savings vary from 11 to 29 per cent depending on the size of investment involved. Having been audited and having received recommendations on to avoid energy wastage, some industrial plants already enjoy the benefit of decreased energy consumption.

In line with the above energy audit promotion, various supporting programmes in the field of energy conservation have been conducted regularly. There is a subprogramme of information dissemination. Among others are bulletin and technical manuals. Another subprogramme is training activities. So far, a Presidential Decree has been issued to promote energy conservation in administrative buildings. Further regulation on compulsory reporting is still being discussed within a working group.

The most recent programme is in institutionalizing energy conservation. The government has created an energy conservation corporation. This government-owned corporation provides consultancy services, training, information and the like.

(iv) Indexation

Measures have been adopted to develop non-exportable/non-tradeable resources of energy or those having a relatively low value in the international market such as geothermal and hydropower to meet the needs of domestic consumption. Efforts have been undertaken to see new possibilities to promote the use of liquified petroleum gas (LPG) as a substitute for kerosene used for household cooking and lighting, and in order to take advantage of the lower economic cost of LPG in relation to kerosene. Another consideration is the possibility of substituting for oil-fired power generation by basing a part of future generation on natural gas. The utilization of compressed natural gas (CNG) in the transport sector (taxis, buses and others) is now under investigation. The natural gas distribution system will be expanded for city gas, since new reserves of natural gas that have been discovered contain less butane and pentane.

(v) Demand management

Energy prices will be studied comprehensively. A study on energy pricing policy funded by the World Bank is now under way. The study will identify methodology and strategy in setting up energy prices that will be of assistance to the Government. At the moment domestic petroleum product prices will be kept unchanged. Coal prices will be established based on the cost of transport.

Export of coal will be promoted in line with the government campaign to increase non-oil exports, particularly of coal from Kalimantan which is of export grade. Measures have been adopted to maintain export of energy as long as this will provide profitable chances and the export of non-oil could not be substituted for it. For this purpose the development of new mines will be enhanced. The production-sharing system is one of the policies in developing domestic resources to attract and mobilize investment.

7. The Philippine energy scenario
(Synopsis)

(a) Supply/demand balance

Since the quadrupling of the world oil prices in 1973-1974, the Philippines has vigorously developed its indigenous energy resources to achieve self-reliance. As a result, the country's dependence on imported oil declined from a high of 92 per cent in 1973 to only 56 per cent in 1986. Indigenous energy production stood at 41.7 million barrels of fuel-oil-equivalent (MMBFOE) in 1986, representing an 86 per cent increase over the 1973 level of only 5.6 MMBFOE. Of the total indigenous supply, nonconventional energy sources, particularly, bagasse and other agricultural wastes, contributed 18 per cent. Other major contributors were hydro and geothermal power, with 11 and 8 per cent, respectively.

Total energy consumption for 1986 was estimated at 94.7 MMBFOE. Of this total, about 40 per cent went to power generation. With regard to sectoral distribution, industry accounted for about 47 per cent: transport, 34 per cent; residential, 10 per cent; and commercial, 9 per cent.

(b) Physical accomplishments

(i) Energy resource development

Cumulative domestic production from the three oilfields in Palawan since 1979 reached 31.8 million barrels as of 1986, with Nido oilfield contributing 48 per cent of the total production.

On the other hand, local production of coal increased significantly from 39,000 MT (metric tons) to 1.2 million MT between 1973 and 1986. Proven reserves totalled 369 million MT in 1986. Meanwhile, coal consumption declined to 1.9 million MT in 1986 from a peak of 2.4 million MT the previous year owing to the stoppage of operations of a nickel mine and refinery and reduced operations of a large copper mine.

The development of the four geothermal fields in the country — Tiwi, Makban, Tongonan and Palimpinon has provided the country with substantial steam availability for its geothermal power generation. Installed plant capacity of the geothermal plant totalled 894 MW as of 1986.

Table 1. Philippine energy mix
(Million barrels of fuel oil equivalent)

	1973		1979		1986	
	Volume	Percentage	Volume	Percentage	Volume	Percentage
Indigenous	5.57	7.98	26.92	27.63	41.71	44.04
I. Conventional	3.32	4.76	13.86	14.23	24.18	25.53
Oil	0.00	0.00	7.18	7.37	2.85	3.01
Coal	0.13	0.19	.82	.84	2.94	3.10
Hydro	3.19	4.57	4.80	4.93	10.47	11.06
Geothermal	0.00	0.00	1.06	1.09	7.92	8.36
II. Non-conventional	2.25	3.22	13.06	13.41	17.53	18.51
Bagasse	2.25	3.22	6.35	6.52	4.09	4.32
Agriwaste		0.00	6.71	6.89	13.01	13.74
Others	a	0.00	–	–	0.43	0.45
Imported energy	64.22	92.02	70.50	72.37	52.99	55.96
Oil	64.22	92.02	70.50	72.37	49.78	52.57
Coal	0.00	0.00	–	–	3.21	3.39
Total energy	69.79	100.00	97.42	100.00	94.70	100.00

Note: a Minimal.

In the nonconventional energy development sector, accomplishments include: (a) generation of baseline data on different types of energy resources used by various economic sectors; (b) identification of availability of nonconventional technologies through the Affiliated Non-con Centers (ANCs) and other institutions and corporations; and (c) encouragement, guidance and/or funding assistance for the installation and monitoring of specific nonconventional technologies such as solar water heaters, biogas digesters, rice-hull or charcoal-fueled gasifier, hydrous alcohol plant using sugar-cane and sweet sorghum, wind turbine, and solar-powered (photovoltaic) pumping system.

In 1986, the share of nonconventional energy resources to total energy consumption was recorded at 18.51 per cent which is equivalent to 17.5 MMBFOE. Of this total agriwaste and bagasse were the major contributors, accounting for 69 and 23 per cent, respectively.

(ii) Power development

In the power sector, the past year saw continued efforts to increase and diversify generating capacity sources and extend supply to new service areas. With the commissioning of four National Power Corporation (NPC) power plants, total installed generating capacity in the country for 1986 reached 6,462 MW, a 4 per cent increase (240.5 MW) over the previous year. In an effort to reach all 56 provinces nationwide, NPC also completed backbone transmission lines spanning a total of 11,974 circuit kilometres and substation capacity of 13.5 billion volt-amperes. These made possible the connection into the system of some private utilities and self-generating industries to avail of more reliable electricity supply.

Total power generation in 1986 amounted to 23,166 GWh, or 38.61 MMBFOE which accounted for 40.8 per cent of the country's total energy consumption. Of this total, oil-based capacity supplied 35.6 per cent while non-oil sources such as hydro, geothermal and coal accounted for 27.1, 20.5, and 11.7 per cent respectively. Of the total power demand, the industrial sector had the largest share of 38.8 per cent, followed by the residential sector with 19.1 per cent and then commercial sector with 14.1 per cent.

(iii) Electrification

The national rural electrification programme provided electricity to an additional 104,000 households this year. To date, the number of households serviced by electric co-operatives has reached 2.8 million, representing 49.1 per cent of the total number of households in the co-operatives' franchise areas. Likewise, a total of 1,270 municipalities and 19,680 barangays are enjoying electricity from 118 registered co-operatives.

The private electric utilities, on the other hand, covered a total of 2.7 million households in their franchise areas. The largest of these utilities, the MECO serviced some 1.9 million households, roughly 91 per cent of the total households in its franchise area. At the end of the year, about 58 per cent of all households in the country had been provided with electrical service.

(c) Outlook

The medium-term energy plan (1987-1992) aims to achieve a higher degree of energy self-reliance, with import dependence to be reduced further from 56 per cent in 1986

Table 2. Energy supply mix
(In million barrels of fuel oil equivalent, MMBFOE)

	1986[a]		1987		1988		1989		1990		1991		1992	
	Volume	Percentage	Volume	Percentage	Volume	Percentage	Volume	Percentage	Volume	Percentage	Volume	Percentage	Volume	Percentage
Indigenous energy	41.71	44.04	45.73	46.88	50.04	48.00	52.90	48.63	54.90	48.10	61.32	50.48	64.43	51.49
I. Conventional	24.18	25.53	25.64	26.28	28.88	27.70	30.53	28.07	31.28	27.41	36.36	29.93	38.00	30.37
Oil	2.85	3.01	1.51	1.55	1.06	1.02	.77	.71	.20	.18	1.62	1.33	1.26	1.01
Coal	2.94	3.10	4.93	5.05	7.43	7.13	8.86	8.15	10.04	8.80	12.29	10.12	14.20	11.35
Hydro	10.47	11.06	11.02	11.30	11.75	11.27	11.93	10.97	12.03	10.54	12.14	9.99	12.14	9.70
Geothermal	7.92	8.36	8.18	8.39	8.64	8.29	8.97	8.25	9.01	7.89	10.31	8.49	10.40	8.31
II. Non-conventional	17.53	18.51	20.09	20.59	21.16	20.30	22.37	20.57	23.62	20.70	24.96	20.55	26.43	21.12
Bagasse	4.09	4.32	4.69	4.81	4.93	4.73	5.20	4.78	5.48	4.80	5.79	4.77	6.13	4.90
Agriwaste	13.01	13.74	14.97	15.35	15.72	15.08	16.58	15.24	17.50	15.33	18.46	15.20	19.57	15.64
Others	.43	.45	.43	.44	.51	.49	.59	.54	.64	.56	.71	.58	.73	.58
Imported energy	52.99	55.96	51.82	53.12	54.21	52.00	55.87	51.37	59.23	51.90	60.16	49.52	60.70	48.51
Oil	49.78	52.57	48.92	50.15	50.48	48.42	53.44	49.13	56.71	49.69	57.81	47.59	57.74	46.14
Coal	3.21	3.39	2.90	2.97	3.73	3.58	2.43	2.23	2.52	2.21	2.35	1.93	2.96	2.37
Total energy	94.70	100.00	97.55	100.00	104.25	100.00	108.77	100.00	114.13	100.00	121.48	100.00	125.13	100.00
Growth rate, per cent P.A.		2.86		3.01		6.87		4.34		4.93		6.44		3.00
Power use, per cent of Total Volume	38.61	40.77	39.37	40.36	43.38	41.61	45.51	41.84	48.94	42.88	51.71	42.57	54.29	43.39
Oil share in power use, per cent	13.74	35.59	12.36	31.39	13.38	30.84	14.80	32.52	17.23	35.21	18.57	35.91	16.87	31.07

Note: [a] Actual.

to only 48 per cent by 1992. The task is not as enormous as in the past decade, though, such that indigenous energy resources development and fuel diversification efforts can now be pursued at a much reduced pace. Over the plan period, total energy demand is projected to grow at an annual average of 4.8 per cent, reaching 125.1 MMBFOE by 1992 from the projected year-end (1987) level of 97.6 MMBFOE. Of this total, coal is planned to supply 11 per cent; hydro 10 per cent and geothermal 8 per cent. Agri-waste and bagasse are still expected to be the major contributor to the indigenous energy supply, accounting for about 40 per cent of the total 1992 volume of 64.4 MMBFOE.

The above targets may yet have to be revised, as certain assumptions used in their formulation already need substantial updating.

B. Energy issues

1. Prospects for production and utilization of coal, natural gas and electricity*
(E/ESCAP/NR.14/10)

(a) The importance of large-scale energy systems in the future

Long-term economic growth is one of the most important policies for the developing countries in the region. Each of the countries hopes (and is making every effort) to eradicate poverty and raise the standard of living for its people.

---

*   Note by the ESCAP secretariat.

For the realization of economic growth, a sufficient and cheap energy supply is one of the basic elements, while people tend to request a more intensive use of energy as their standard of living rises with economic growth. As a result, the increase in energy demand continues. According to research conducted by an ESCAP consultant, even with strong conservation efforts energy demand for developing countries in the region will increase 2 to 2.5 times by the year 2000 and 3 to 3.5 times by the year 2010 as compared with demand in 1983 (table 1). Enlargement of large-scale energy supply systems is necessary to cope with the predicted increase in energy demand.

Among various energy resources, it is expected that demand for oil will still be large, although price levels will influence it. The biggest merit of oil is the maturity of its system and technology of production, supply, usage, and so on (for example, exploration activities are being carried out in a systematic manner; production techniques in severe conditions such as the Arctic, or in the sea, have been well established; there are pipeline networks and convoys of tankers; refinery facilities exist all over the world as well as a world-wide oil market, gas station networks, techniques of gasoline engines, and so on). The existing mature system and technology may be able to increase the supply of oil through a gradual enlargement of the system itself and through the progress of technology. At present, there is a wealth of manpower and know-how related to oil which has been accumulated by the oil industry. Additional investment for enlarging these factors may be relatively small and the risk of investment is also not very high, while supplies are available. Considering the advantages of this system, a larger oil supply system may not be too difficult to achieve.

Table 1. Estimates of Asian energy and oil demand
(Million tons of oil equivalent)

	Energy[a]			Oil		
	1983	*2000*	*2010*	*1983*	*2000*	*2010*
	Actual			Actual		
Industry	82	175-197	257-289	28	58-63	128-141
Transport	53	101-192	139-208	47	91-182	125-187
Residential/commercial	38	69-83	124-149	20	36-43	62-73
Power	–	–	–	26	34-56	60-100
Total	173	345-472	520-646	120	219-344	375-501

Source: "Energy demand in Asian development countries: historical trends and future prospects" submitted to the Asian Forum on Energy Policy, 27-30 October 1986, by Mr. Jayant Sathaye of Lawrence Berkeley Laboratory, California, United States of America.

[a]   Final demand; includes electricity.

Note: Countries included in this forecast are Bangladesh, India, Indonesia, Pakistan, Philippines, Republic of Korea and Thailand. In addition to the above countries, Mr. Wu Zhong-Xin at Tsing Hua University estimated a similar energy demand increase for China, at an average rate of 4.4 per cent through 2010.

Assumptions: These estimates are based on estimates by a consultant as to the overall energy intensity and oil intensity of each of the major sectors, and expected changes in each sector.

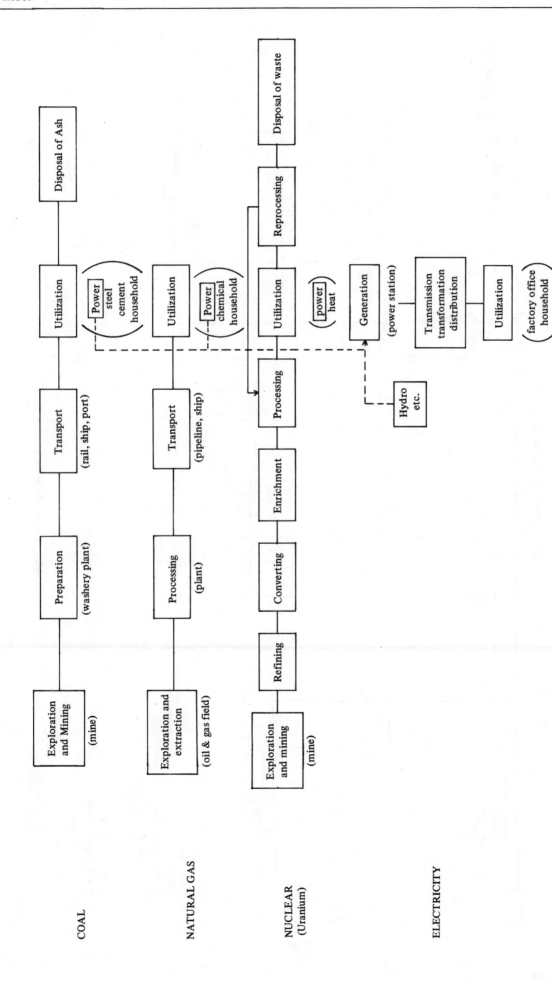

Figure I. Various systems of energy supply

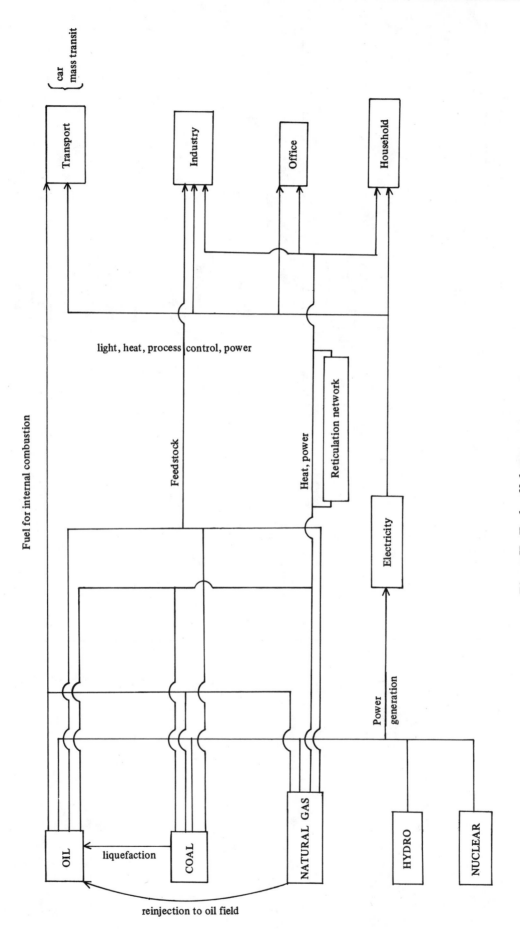

Figure II. Trade-offs between resources

Availability of oil resources is, however, a problem. During and just after the oil crises, the limitation of oil supply was a big problem, and in addition, as many experts predicted, a long-term problem. Recently, owing to conservation, substitution and so on, the price of oil has not been as high as before. It seems that at present oil is available at a "reasonable" price. But this situation should not lead to complacency. As stated earlier, the long-term demand for energy and oil is vast even for the selected countries of the ESCAP region, while the present size of proved world reserves is only 143.7 billion tons of coal equivalent (34.4 years of reserves/production ratio at current consumption levels)[1].

After the oil price shock, most countries have been promoting conservation and substitution programmes, but under the present depressed market conditions for oil, such activities seem to be lagging. In order to avoid (or alleviate) any long-term bad effects from oil resource limitations, in addition to intensive efforts for energy conservation, research into and development of substitutes for oil, such as coal, natural gas and electricity, have to be strengthened considering their resource availability position.[1] For a large increase in the supply of coal, natural gas and electricity, the establishment of (or considerable enlargement of) complete systems, from exploration, production, transport to utilization, is necessary (figure I). Without harmonized chains of systems, smooth delivery of vast amounts of energy resources cannot be realized. As each of the systems requires considerable financial and human resources for many decades, it is necessary to consider and establish firm plans and ideas on the degress of necessity (and difficulty involved) in strengthening these systems. This can be done through studies on the trade-offs of various energy resources and on the techno-economic aspects of each component of the systems (figure II).

(b) The role of coal

The world-wide availability of coal is vast compared with other energy sources (six times more than oil). In

addition, the occurrence of deposits is more dispersed than deposits of oil and natural gas. Owing to this availability and accessibility, coal may have a considerable potential for economic development. It can be used in various ways: it is used primarily for power generation, and is also used extensively for the smelting of iron ore, and for cement production. Industrial fuels, chemical feedstock, and coal gas and briquettes for household use are other forms of intensive usage.

As there is a relatively low endowment of oil resources, the region expects a large role for coal in the future energy supply. Even now, a vast amount of coal is used in Australia, China and India. The last two countries will increase coal production and utilization sharply. Other countries are also promoting the introduction of coal. As stated earlier, power generation may be the main use. Wherever introduction into power system use occurs, large systems covering all areas from mining to power station, and, finally, ash disposal, have to be established.

Some countries have embarked on the use of imported coal. A power station based on overseas coal may entail the most complicated logistics system. Establishing such a coal chain involves, not only construction of mines and power stations but also the establishment or enlargement of inland transport facilities, ocean transport, loading and unloading ports, and so on. The power station may rely on several supply sources. It therefore also has to install facilities for storage and mixing of various brands of coal. Development of international coal markets may be envisaged. Japan and the Republic of Korea already have such power stations. Malaysia, the Philippines, Sri Lanka and Thailand, and possibly other countries of the ESCAP region, are also promoting such a project.

Coal has several disadvantages. As a solid, handling it is more difficult than handling liquid fuels. The transport of bulky cargoes is not easy. In the case of China and India, where coal reserves are concentrated in a few regions while the demand is dispersed all over the country, the transport problem is a major constraint on the wider utilization of coal. There are various environmental problems, such as the distruction of the environment around mining

---

[1] For resource position, please see tables 2 and 3 in article 1 *the regional energy scene,* chapter A.

	Available	*Intermediate (around 2000)*	*Future (around 2010)*
Mining		Greater mechanization	
Preparation			
Transport		methanol	Gasification, liquefaction, Coal-water mixture
Combustion	Fluidized bed combustion		

In situ gasification may be introduced later than 2010.

Figure III. Utilization possibilities for coal-related technologies

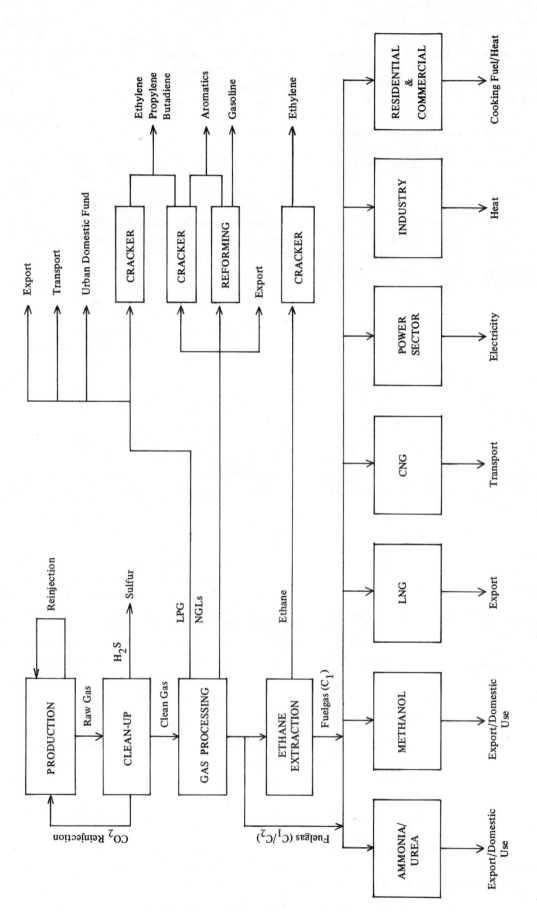

Source: ESCAP, *Techno-economic Study of Natural Gas Production and Use* (ST/ESCAP/455).

Figure IV. Summary schematic of gas processing options

areas, pollution with dust/ash particles, SO_x, NO_x, "acid rain" due to combustion, etc. For intensive use to become possible, efforts on technological innovation should be strongly promoted to avoid any possible damage.

For a safe and cheap means of production, the mechanization of mining should be pursued. However, even with the general increase in mechanized coal mines, a large number of small mines will continue to operate with traditional, labour-intensive technologies, with a view to supporting local economic and employment interests.

As an effective use of low-quality coal, fluidized bed combustion systems have been introduced for commercial generation in China and India, among others. Emission-control techniques are also being developed intensively, and in the case of a newly-constructed power plant, over 99.7 per cent of the particles can be contained.

For long-term prospects, gasification, liquefaction and coal-water mixtures may be important, as they can remove some or most of the bottlenecks. Regarding technologies, gasification and liquefaction have been established previously, but are still not always economical. In the long run, oil from coal may be important.

(c) Role of natural gas

Resources of natural gas are not as large as those of coal, but occurrences of deposits are more dispersed than oil. Natural gas can substitute for oil in various fields of utilization (fuel, chemical feedstock, CNG used for vehicles, small amounts of LPG extracted from natural gas for domestic and transport use, and so on). All natural gas producing countries therefore include the utilization of indigenous natural gas as one of the key factors in their energy policies.

Natural gas can be utilized in many ways (figure IV), and therefore trade-offs between various options have to be carefully studied. It is noted that the various means of utilization differ among countries, but they share the common objective that natural gas should be used for improving their balance of payments and that natural gas resources should be preserved for long-term optimum use. At present, a large amount of natural gas is being used for power generation, but all the natural gas producing countries plan to limit the usage for power generation after a while, for other more "noble uses". Some countries have already initiated such measures. City gas through pipelines is available only in limited areas. In the long run, many more big cities in the region may introduce city gas systems.

The development of a natural gas project is generally speaking more risky than other projects. It is site-specific. As there is not a world-wide market, a closed circuit chain from the production source to the user has to be established. The necessary funding is very large (in the case of an LNG chain, initial investments of $US 3.5 to 6.0 billion by exporters is quite common). Careful studies on the total system in its techno-economic aspects should be made in advance of the development. During the period of high oil prices, even such a large investment might not have looked so risky, but as the price of oil dropped the impetus for the development of large-scale natural gas projects has dwindled.

As the occurrences of deposits are similar, natural gas and oil can be considered as one group, especially for prospecting and exploration. Recent oil price reduction slowed exploration efforts, and may lead to earlier supply shortage of both oil and natural gas. For the purpose of more intensive exploration and development, government financial aid, or various incentives for investment, are desirable, and can in many instances be negotiated on a case-by-case basis.

(d) Role of electricity

It is expected that in the future vast amounts of energy will be used everywhere and therefore systems will have to be established to meet such demand: oil and electricity are possible systems. As oil resources are limited, for long-term development an electricity supply system can best fulfil such requirements. As already stated, coal and natural gas can be used as fuels for power generation, and can be integrated into the electricity system.

Electricity has various uses in industry as well as in households, and is indispensable as the energy source for communications, data processing, controlling of automation and so on in a high-technology society. Electricity also helps people to enjoy a convenient way of life. Demand for electrification in rural areas is very strong. As economies grow, the quantity demanded will increase considerably, and the quality requested will become higher.

It may be mentioned that electricity supply systems have already been developed in all the big cities in the region, and that rural electrification is proceeding. This advanced state of development of a system may constitute the big advantage of electricity. However, enlargement and improvement of the system in an evolutionary manner are still necessary for future increasing demand. For sources of electricity generation, reference to coal and natural gas has been made earlier. Hydroelectric power and nuclear energy are other main sources of power generation.

There is much unharnessed hydroelectric potential in the big rivers. Such power will have to be developed vigorously, and in concert with environmental control and the overall development of the river, including irrigation projects, flood control projects, transport use, and so on.

As big hydropower stations are usually located far from the demand site, long-distance transmission lines and big substations for transformation near the demand site are indispensable. As electricity is generated and consumed simultaneously, the system of generation-transmission-transformation has to be prepared and operated as one unit. The initial investment for establishment of this system is therefore very large.

Nuclear energy is a newly-developed form of energy resource as compared with other conventional energy sources. A vast supply of energy can be obtained through a nuclear reactor. As such reactors require a relatively small amount of fuel for energy, this fuel supply does not burden a country's balance of payments as much as oil and coal, another obvious potential advantage.

Harnessing nuclear energy, however, requires big power systems with high technology and a high standard of safety. Before embarking on an intensive use of nuclear power generation, industrial support capability and trained manpower have to be available. Complicated nuclear fuel cycles, including enrichment, reprocessing, disposal and so on, also have to be established, at least in some countries. As a result, the number of countries to utilize nuclear power intensively will be limited. For any progress in the field of nuclear utilization, international co-operation in various aspects, such as technology, safety, financing and so on, is essential.

In addition to the supply of a sufficient volume of electricity, improvement in the quality of electricity (for example, stabilization of frequency and distribution reliability) is also necessary. Such improvement requires strengthening of networks. This applies especially to metropolitan areas, where high demand and the use of sophisticated equipment will be concentrated; big network loops surrounding the cities (to ensure a supply of high-quality electricity) may have to be developed. One way of strengthening networks is through trans-country power networks. These networks will have a considerable advantage for all the related subregions, through the improvement of supply reliability and possible joint developments in power potential.

These network enlargement programmes of the power supply system will call for a vast amount of investment. As most developing countries suffer from lack of funds, especially foreign exchange for imports, financial co-operation by developed countries and aid agencies is more essential than before. Developing countries by themselves have to establish their capability to raise large funds internally and internationally. In this regard, the financial viability of power supply agencies is very important. Efforts to improve operational efficiency through reduction of generation, transmission and distribution losses have to be pursued effectively. Adequate pricing systems to cover electricity costs and to improve the internal fund-generating capacity for power supply agencies should be sought, with due consideration to possible electricity price subsidy effects on supply, and also the welfare of the poor.

(e)   Summary and recommendations

For long-term economic growth, oil resources may not be sufficient to supply the necessary quantity of

energy. On the other hand, as the growth of economies continues, a better quality of energy will be demanded. To cope with these requirements, intensive development of coal, natural gas and electricity systems from production to utilization will have to be undertaken.

Coal resources should be developed mainly for power generation, as the availability of coal and technological adaptability render it generally acceptable for such use as compared with other uses. However, for intensive utilization, various technical innovations should be strongly pursued so as to avoid possible environmental problems.

For the development of natural gas resources, careful studies in advance on the trade-offs of various ways of utilization, and techno-economic studies on total system design should be carried out to find the optimum way of utilization, as natural gas can be used for various ends; development of natural gas resources is influenced by individual conditions, including location, far more than in the case of other energy resources. In principle, natural gas may be phased later into uses such as city gas, through a reticulation network, and chemical feedstocks once the infrastructure investments are in place. For economic reasons, early uses will invariably be for power generation. To increase exploration, new incentives are desirable, especially under the current lower oil price regimes.

As a convenient form of energy, electricity supply networks should be strengthened, in order to cope with the need to raise the standards of living for both rural and urban residents. Considering the potential of nuclear energy, studies on a long-term development strategy in developing countries should be continued. For realization of the enlargement of the power supply systems, the financial viability of power supply agencies has to be strengthened.

For the development of an optimum mixture of these various systems, a careful study of trade-offs in advance is necessary, as each system is expected to exist for decades.

To organize and run these systems, the mobilization of various large resources (such as energy resources, funds, technology and know-how, manpower, etc.) as well as long-term commitment for the system is required. In this respect, intensive international co-operation, such as joint investment and joint operation of projects, long-term sale and purchase contracts, and exchange of techniques and knowledge, and assistance, is highly recommended. ESCAP stands ready to assist in facilitating such intercountry arrangements at the technical level through the consulting mechanism of subject-specific TCDC working groups, expected to come into existence gradually in 1987-1991 through the regional energy programmes.

2. The use of coal in households and small-scale industries*

Introduction

In phase II of the first cycle (1984-1986) of the UNDP-financed regional energy development programme (REDP) for the Governments of the developing countries of Asia, ESCAP undertook to study the potential for coal and peat use in small-scale industries and rural households in countries with dispersed coal deposits.

The study design incorporated TCDC (technical co-operation among developing countries) elements from the very beginning: consultants from developing countries with measurable progress in the field of coal utilization in households and small-scale industries were to advise researchers and government officials in interested countries having available coal resources.

The result of the study was largely negative: as long as coal or peat briquettes compete with non-commercial fuel-wood, very little penetration into the rural household market can be expected.

The TCDC approach has proved its value not necessarily in the way the design of the study envisaged it: the negative experience of some "advanced" developing countries in the field of coal utilization may help some others contemplating large programmes of substitution for non-commercial wood-fuel (for ecological reasons or for reasons of long-term sustainable commercial energy supplies) to avoid costly mistakes related to possibly ill-conceived product designs and/or pricing policies.

Some countries (Republic of Korea, Viet Nam) have, in fact, achieved commendable penetration rates of briquetted coal and biomass into the "near-commercial" markets, and as a result have become possible examples to follow. The importance of local solutions to local problems through dispersed testing facilities was emphasized in the study.

Lessons on non-commercial fuel supplies also have a philosophical dimension: should the problem be solved by an abundance of non-commercial fuels (if technically feasible and economically affordable for the country) thus perpetrating rural subsistence-level income patterns, or should one aim at first increasing productive income levels and then supplying high-grade energy (electricity, LPG, etc.) later? What happens in the transition period? Is the right problem being solved?

The lessons learned from the six ESCAP country studies are valuable for having raised these questions once again, while at the same time pointing to the fact that categorization into "fuelwood and charcoal", "coal briquettes", "kerosene", "LPG", and "electricity" market segments

_____

* Note by the ESCAP secretariat.

sometimes hides the fact that all these are alternative (more or less attractive) methods for filling basic energy needs, and should be analysed in a dynamic, evolving, integrated context.

(a) Pricing and marketability issues in six countries

At the project team meeting on the above-mentioned coal utilization study held from 6 to 8 May 1986; the information presented about the data on India, the Philippines, Thailand and Viet Nam was reviewed. Data from Indonesia and the Republic of Korea were subsequently reviewed outside the project team meeting. The main conclusions are given below.

(i) Data on India

In reviewing the Indian data, the long experience with different briquetting techniques was emphasized. In spite of this, only about 3 per cent of coal use (about 4.5 million tons per annum) is in the domestic sector, vis a vis 47 per cent for direct burning in large- and medium-scale industries (steel, cement, fertilizer, paper, glass and finally, rail transport) and 50 per cent for electric power generation of the 150 million tons of coal produced annually. Competition with "free" firewood was identified as the main obstacle to penetration in the rural domestic sector. Consumer acceptance of briquettes made out of "washery sinks" and "coke breeze" in the coal-producing regions in northern and eastern India has led to the establishment of a large number of units catering for this demand.

(ii) Data on Indonesia

The obstacles to introducing coal as an alternative energy for small-scale industry are as follows:

(i) Small-scale industries generally use traditional methods for fuelwood or charcoal combustion (technical obstacle);

(ii) Unreliable continuous supply of coal from local small mines, or for the small-scale user (distribution obstacle);

(iii) The cost of coal transport is high, rendering production costs uncompetitively high for the users (market obstacle).

In spite of this, the role of coal as an alternative non-oil energy source became important in recent years based on the Presidential Instruction of 16 September 1976 stipulating that the development of coal should replace the role of oil in the economy.

A survey conducted by the Center for Development of Mineral Technology (Pusat Pengembangan Teknologi Mineral (PPTM)) has indicated that the coal potential in Java is approximately 34.7 million tons of steam-coal quality, with a calorific value of 4,000-7,000 kilocalories per kg. PPTM has established 18 pilot projects for small industries for coal utilization, using local coal reserves. The coal potential in South Sulawesi is estimated at 39.9 million tons, in Irian Jaya, 4.0 million tons. (Of course, in addition to this, large deposits of export quality coal exist in Kalimantan, Sumatra and Irian Jaya, estimated at over 12 billion tons).

(iii) Data on the Philippines

During the review of the Philippine experience, it was emphasized that although private sector initiatives have started with briquetting applications, these were stopped by the shortage of domestic coals created by the Government's very successful industrial coal for oil substitution policy. (The only exception is in Negros Island in central Philippines, where a coal-firing alcohol plant briquettes the fines (−1/4″) of coal supplied by various small-scale coal mines to maximize its coal utilization. The briquetting capacity is 3,000 tons per month). For the future, the main briquetting potential is in the upgrading of lignite in northern Philippines via briquetting. This could be done in conjunction with drying, to make transport of the lignite to users in Manila viable, and also to supply small-scale industries nearby (especially tobacco flue curing, as an alternative to firewood). This will produce 225,000 tons of briquettes per annum.

(iv) Data on the Republic of Korea

In 1985, approximately 80 per cent of the homes in the Republic of Korea were heated with briquettes of anthracite coal from domestic and foreign mines. Each briquette, in the form of a cylinder, measures 150 mm in diameter, 142 mm high, and weighs 3.6 kg., 22 holes 14 mm in diameter, as standard specification, and is punched through the cylinder from the top to bottom. This allows for more uniform burning and each cylindrical briquette burns for approximately 10 hours.

Some problems have developed in the manufacture of these briquettes. The domestic coal gradually became poor in quality, necessitating the importation of ever larger tonnages of coal (that is, of higher quality) in order to maintain a certain heating value in the briquettes. There was also a lack of uniformity of heating value of the briquettes because of the variability of the quality of the coal, both foreign and domestic.

This was no small matter, because over 5 billion briquettes are manufactured annually at the 260 plants in the Republic of Korea. About 18.5 million tons of anthracite coals are supplied from domestic mines, and approximately 2 million metric tons are imported from foreign countries.

Basically, there were two problems. First, the Government was concerned with the high variability of the heat content of the coal briquettes and has decreed that the average heating value of briquettes be 4,600 kilocalories/kg and the minimum value 4,400.

Second, maximization of the use of domestic coal to make the briquettes was desired by all concerned. The imported coal has a range of heating values from approximately 4,900 to 6,000 kilocalories/kg and is mixed with domestic coal (normally of a lower heating value) in order to bring the average value up to the acceptable standard.

These problems have been solved successfully through the work of the Korea Institute of Energy Resources and thus the distribution and marketing system for coal briquettes continues to function properly; about 20 million tons of briquettes are sold annually (leaving about 10 million tons of residual ash).

(v) Data on Thailand

(i) Although there is a long history of experimenting with lignite briquettes and a large unsatisfied demand, especially in the eight provinces around Bangkok, there has so far been no real technical breakthrough in satisfying this demand;

(ii) Obstacles to market penetration (in addition to technical ones) include high transport costs and the current low oil prices (competition from kerosene and LPG);

(iii) A Japanese pilot project soon under way might help to overcome the technical obstacles;

(iv) Competitive pricing is important, to permit commercialization.

(vi) Data on Viet Nam

Although Viet Nam has various energy sources (e.g. oil, natural gas, hydroelectric power, etc.), coal is the main energy source for industrial and household use, in both urban and rural areas. Anthracite reserves are around 3 billion tons, but there are also other grades of coal, some of it of coking quality, steam coals and coals with high volatile content, such as lignite and peat. Coking coal reserves are estimated at 15 million tons.

Transport of coal from places of production to large end-users and also to storage yards of the regional coal supply stations that handle small users takes place within the purview of the Coal Supply Corporation under the Ministry of Mines. Transport is arranged mainly by waterways and rail for long distances and by road for short distances. Regional coal supply companies then deliver coal to briquette makers to produce briquettes for household consumption at fixed local prices. The volume is estimated at about one million tons a year, in addition to the industrial quality briquettes manufactured centrally for the use of railways and large industries at a rate of 60-70,000 tons per year.

(b) Integrated planning imperatives

Why is integrated energy planning important? An excellent illustration can be given from Ramesh Bhatia's analysis of product, prices and production of soft coke in India.

"Soft coke, as produced and marketed in India, is manufactured from coals with some coking properties. It is used for household cooking, as well as for input in brick kilns. Production of soft coke by Coal India Limited has been declining from 3.25 million tons in 1976-1977 to 2.41 million tons in 1979-1980 and to 1.74 million tons in 1982-1983. This lower availability and use of soft coke would have resulted in high consumption of kerosene and/or fuelwood in cities and small towns. This reduction in the output of soft coke has been due to a variety of factors including:

"1. Though kerosene is subsidized by the government up to 25 per cent of its c.i.f. price, the subsidy on soft coke has been only US$ 4 per ton. Even the effect of this subsidy is partly eroded by the royalties/cesses attracted by coal (used as input in manufacture of soft coke), mainly from the state governments.

"2. At present market prices, soft coke is at a considerable economic disadvantage with respect to kerosene and LPG because its calorific value is about 60 per cent of that of kerosene or LPG, and the efficiency in use, or the appliance efficiency in the case of soft coke, is only about 20 to 25 per cent compared with 50 per cent and above in the case of kerosene and LPG. This means that, although in terms of Rs/kg soft coke is cheaper than kerosene and LPG, in terms of Rs/kcal (Rs/kg divided by kcal/kg) and Rs per effective kcal (Rs/kg divided by kcal/kg multiplied by appliance efficiency) kerosene and LPG are cheaper than soft coke. Even when LPG and kerosene are valued at import parity prices, soft coke has only a marginal advantage, and that too in specific locations in the eastern region. Besides, it is more convenient to use kerosene/LPG devices as these can be turned on and off whenever the consumer requires it, and the flame is of uniform intensity.

"3. Transportation cost is a major component in the market price of soft coke. Average transport costs have been increasing over time as the dispatches by rail have declined from 1.4 million tons in 1976-1977 to 0.54 million tons in 1982-1983.

"4. The quality of soft coke has declined over time since there are no differential prices based on quality. If overall profitability is the criterion, the local management in coal mines tends to ignore quality if that helps to improve the price of run-of-mine coal by reducing coal allocation for coke making.

"5. Since soft coke has not been given the same level of subsidy as that given to kerosene, the producers have not been getting remunerative prices.

According to a recent estimate, the producer will incur a loss of $US 2-10 for every ton (on a soft coke price of $US 17.5 fixed by the government), depending upon the type of coking coal used as input.

"6. The result of low prices by the government has been that soft coke has been produced in a traditional manner without any control on quality. Investment in large modern plants has not been made because of low anticipated demand which, in turn, is due to high consumer prices and low quality of soft coke. For example, the Government of India recently turned down an investment proposal for manufacturing one million tons of soft coke involving a total capital cost of $US 24.2 million. This investment was not considered attractive even when the pay-back period was seven years at the existing controlled price of soft coke. The Bureau of Industrial Costs and Prices of the Government of India has recently recommended that urgent consideration be given to the one million ton plant – with an increased subsidy of $US 10 per ton for the consumer – so that the market can be flooded with soft coke. With the higher subsidy, the market price of soft coke would come down to $US 11 per ton so that more people might be inclined to switch back to soft coke from kerosene (which is imported, involving foreign exchange). Further, a reduction in the price of soft coke would influence consumers of firewood and charcoal not to switch to kerosene as their incomes increase but instead continue to use soft coke because of the price differential. A lower price of soft coke would also stem the flow of firewood from rural areas to the cities.

"7. A remunerative price of soft coke would also encourage investment in manufacture of high-quality (smokeless) soft coke and town gas by using a low- or medium-temperature carbonization method. This smokeless soft coke would induce people back from the use of kerosene to soft coke.

"Thus, the case of soft coke in India illustrates the following aspects of pricing policies in the energy field:

"(i) Distortions were introduced in the consumption pattern when the price of one fuel (kerosene) was subsidized, while that of a close substitute (soft coke) was not subsidized (or at least not subsidized to the same extent). This resulted in the shift from soft coke to kerosene involving a reduction in demand and output by 5.6 million tons in a 5-year period from 1977 to 1982. Assuming that this reduction in availability of soft coke was replaced by imported kerosene, this would have resulted in an increase in kerosene imports by 1.56 million tons. The estimated cost of these kerosene imports was on the order of $US 450 million over a 5-year period, an expenditure which could be easily avoided.

"(ii) The price paid to the producer was not made remunerative enough to improve the quality of soft coke and to make investments in new units. For example, the price obtained by the producer was not distinguished by grade so that there was no incentive for the producer to maintain the quality of coals used for soft coke. The result was bad quality coals resulting in low-quality soft coke, reducing its demand further. There was also no incentive to modernize the plants used for manufacturing soft coke in order to improve the quality of soft coke and recover tars. Besides, investment in modern units and Low Temperature Combustion plants was not allowed since it was feared that there would not be adequate demand for soft coke.* In this way, an unimaginative pricing policy for soft coke resulted in a vicious circle of low demand, lower quality; low investment, low output, and lower supply. This, in turn, resulted in a foreign exchange outflow of $US 450 million for kerosene imports over a five-year period."[1]

In Indonesia, where a large-scale coal substitution policy in industry (cement, etc.) is promoted, this could result in domestic and small industrial customers for coal possibly having lower priority in the eyes of mine operators than the often more profitable large-scale industrial customers; at the same time, as kerosene is a (subsidized) domestic fuel, pricing problems of a similar nature might very well be expected to arise. Looking at the Indian and Philippine experience, TCDC in this case could well lead to the avoidance of possible errors in the Indonesian coal utilization programme, and thus the fact that an Indian TCDC consultant was involved with the Indonesian study may well have served a very useful purpose. A similarly useful purpose may have been served by the Philippine TCDC consultant's involvement in the Thai study.

---

* Note that only if best quality soft coke produced through new investments would be at an economic disadvantage compared with unsubsidized kerosene, then and only then would the foreign exchange outflow be justified since customers would prefer a better-grade fuel at a lower price in buying kerosene, *ceteris paribus*.

[1] Ramesh Bhatia: "Energy pricing in developing countries: role of prices in investment allocation and consumer choices", C.M. Siddayao, ed., *Criteria for Energy Pricing Policy*, (London, Graham and Trotman, 1985), chap. 5.

Lessons learned through TCDC from the Republic of Korea and Viet Nam are of a different nature: rather than the avoidance of errors, there exist positive examples worthy of imitation, specifically on the consumer acceptance side.

Annex I gives the Vietnamese information, while annex II gives information on coal briquetting in the Republic of Korea.

The lessons learned from both of these is that as long as the fuel requirements of end-users are taken into account in designing the final fuel product, appreciable market penetration into the household and domestic market is possible. In Viet Nam, this design involves shorter burning times mainly for cooking applications, with continued reliance on fuelwood for the start-up process. In the Republic of Korea, space heating applications (in combination with cooking in the "ondol" system) require 10-12 hours continuous burning of the briquettes (2 briquettes per day). The large penetration (80 per cent of households) is explained both by the unavailability of fuelwood and an efficient distribution system organized at prices that have taken rural budget constraints into account.

In Viet Nam, the penetration rate is comparatively high, in spite of the availability of fuelwood, not only because of the well-organized distribution system but also owing to product design taking locally available materials (agricultural residues, peat) into account. This system requires a network of testing laboratories with measuring equipment and qualified personnel. One of the recommendations of the visiting TCDC consultant was that such equipment might be provided through concessionary aid on a TCDC basis as a follow-up activity of the current study.

It may be suggested that the extensive testing experience of the Republic of Korea of anthracite briquettes and blending be made available on a TCDC basis to other developing countries. Testing equipment and possibly briquette plants could also be offered.

As a general conclusion, the TCDC design of the study has achieved its purpose in drawing attention to both the constraints (pricing, coal availability limits, product design, etc.) and the achievements in the utilization of coal briquettes for the domestic and small-scale industrial sector, supplying some examples to follow, and pointing to sources of expertise.

Annex I

THE POSSIBILITY OF USING COAL AND PEAT IN SMALL INDUSTRIES AND HOUSEHOLDS IN VIET NAM

(A synopsis)

I. *Coal characteristics:*

Quality of various coals in Viet Nam

Ash content	5-40	10-30	8-12	7-54	per cent
Volatile	5-7	12-24	40	15-54	per cent
Sulphur	0.5	1.3	0.5	0.1-4	per cent
Calorific value	7,400-8,500	5,500-7,500	6,600	1,500-5,500	kcal/kg
Name	Anthracite	Coking coal	Others		

II. *Briquette-making experience (industrial briquettes)*

Production formula: 72 anthracite dust
(percentage composition)

20 coking coal

8 coal tar pitch

Total   100 per cent

III. *Coal tar pitch characteristics:*

Melting temperature	Components (percentage)			
°C	A	B	C	*Total*
86	29.94	55	13.5	98.44
76	25.0	30.8	42.6	98.4
72.5	23.1	51.3	24.2	98.6
70.5	22.8	34.8	40.3	97.9

Legend A = Cokes at 500°C, black, does not soften or melt

B = Softens at 90°C and catches fire at 110°C

C = Soft, bluish, flows at 10-20°C

IV. *Standard (railway fuel)*   (70-80,000 tons per year) *briquette quality:*

(with synthetic binder developed in Viet Nam, containing lime and bitumen, hot process tar-pitch binder)

Ash content	less than 18 per cent
Volatile	greater than or equal to 14 per cent
Sulphur	less than 0.5 per cent
Calorific value	7,000 + 200 kcal/kg

V.   *Other kinds of briquettes* (cold process, clay binder)

Various kinds of briquettes are produced in different areas of Viet Nam for *household use* (with various raw materials; percentages)

1	2	3	4
50 anthracite dust	60 coconut husk	65 coal dust (partially burnt coal)	40 coal dust (partially burnt coal)
25 peat	30 coconut waste (after fiber extracted)	25 anthracite dust	35 anthracite dust
10 charcoal	10 binder	10 binder	25 local peat
10 coconut hust			
5 binder			

VI.   *Briquette characteristics* (household use)

	honey comb	"oval"	column
diameter	110-120 mm	35 x 60 x 70 mm	40 mm
height	100-110 mm	38 x 54 x 74 mm	65 mm
weight	1.0-1.2 kg	90 grammes	45-60 grammes
burning time	3-4 hours	1-2 hours	few minutes (shortest)
time till self-sustained burning	15-18 minutes	15-20 minutes	13-20 minutes

Self-sustained burning is achieved by using 150-200 grammes of firewood at the start of the fire.

VII.   *Vietnamese metallurgical coke characteristics:*
       (components, percentage)

Vitrinit	94.5
Cexinit	
Fluxinit	1
Somifuzinit	
Micrinit	1.2
Other mineral matter	3.5

VIII.   *Coal gasification – formation gas characteristics*
        (component, percentage)

CO_2	3.9	
$C_n H_n$	0.2	(n > 1)
CO	27.8	
H_2	15.5	
CH_4	0.8	
N_2	51.5	

Coal is crushed to 15-35 mm or 35-50 mm or 50-80 mm size, produces 4.15 m³ gas per kg of coal with a calorific value of 1,175-1,300 kcal/m³

IX.   National coal organization structure

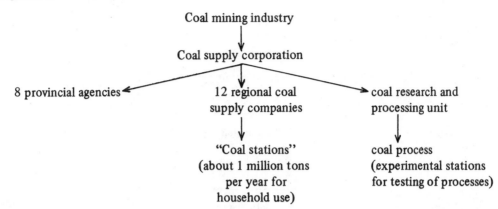

X.   Summary

Although Viet Nam has various energy sources (e.g. oil, natural gas, hydroelectric power, etc.) coal is the main energy source for industry and household use, in both urban areas and the countryside. Anthracite reserves are around 3 billion tons, in Quang minh mainly, but in other provinces, such as Bac thai, Hoang lien son, Lai chau, Ha nam ninh, Nghe tinh, Quang nam and Da nang, there is also coal, some of it of coking quality; others are steam coals and coals with high volatile content, lignite, and peat (U-minh area). Coking coal is estimated at 15 million tons. The structure of the industry is shown in figure I.

As shown in the graph in section IX above, the State organizes transport of coal from places of production to large end-users and also to storage yards of the regional coal supply stations that handle small users through the coal supply corporation under the Ministry of Mines. Transport is arranged mainly by waterways and rail for long distances and by road for short distances. Then, regional coal supply companies deliver coal to briquette makers to produce briquettes for household consumption (figure III) at fixed local prices. The volume is estimated at about one million tons a year, in addition to the industrial quality briquettes manufactured centrally by a hot process (figure II) for the use of the railways and large industries, in the volume of 60-70,000 tons per year.

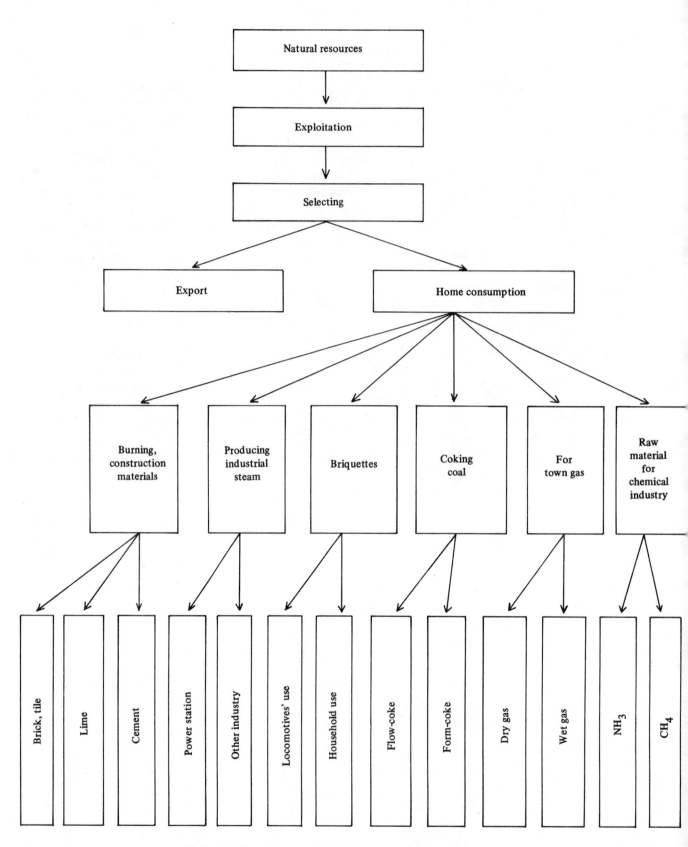

Figure I. The structure of the Vietnamese coal industry

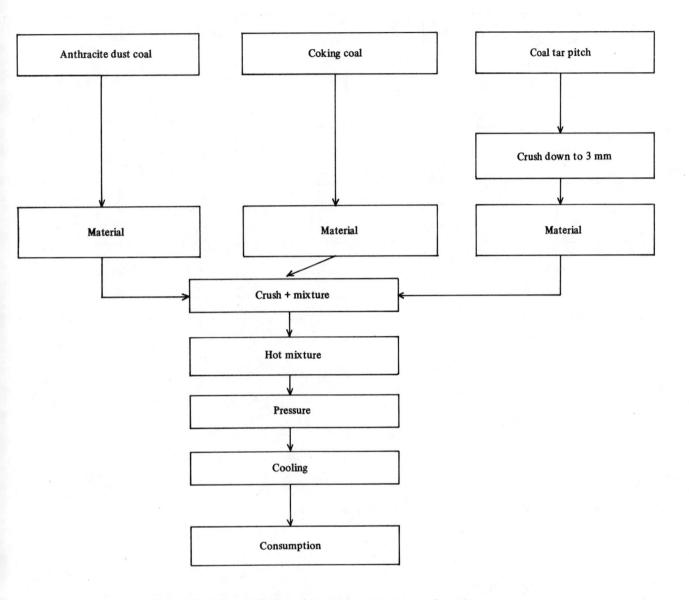

Figure II. Industrial/railway/standard briquette manufacturing process

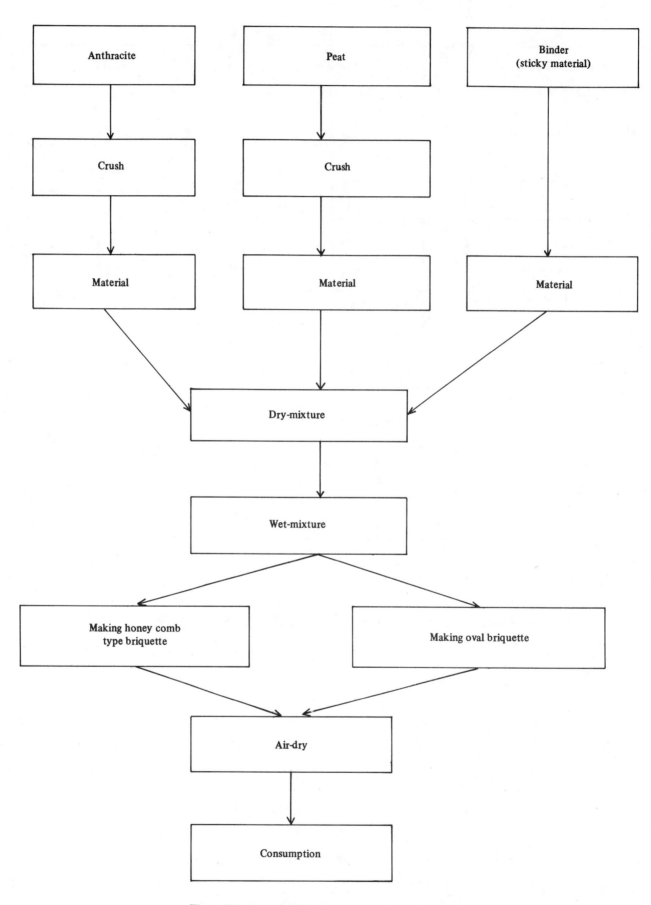

Figure III. Household briquette manufacturing process

Annex II

COAL BRIQUETTE INDUSTRY IN THE REPUBLIC OF KOREA*

In 1980, there were 261 coal briquette plants in operation in Korea, which processed approx. 20 million tons of anthracite coal annually. Table 1 shows the heating value, volatile matter, carbon and ash content of some typical feed coals.

Their capacities ranged from 50,000 tons of coal per year to 800,000 tons per year. Plant capacity and distribution of plant location is showed in Table 2. Standard deviation of the heating value of coals handled by a sample plant is given in Table 3.

The major units of coal briquette plant include pre-blending, size reduction, mixing that improve the quality of different kinds of coals, as delivered from mine to make it suitable for specific standard briquettes.

---

* Reproduced from *Proceedings – Coal Utilization Technology: 1st Korea-USA Joint Workshop, October 11-13, 1984,* Daejeon, Republic of Korea (Pittsburgh Energy Technology Center/ Korea Institute of Energy and Resources).

These operations include sizing; removal of rock and other extraneous materials from the mine roof or floor; crushing to reduce bulk coal to meet proper size of particles to briquette; blending of coal from different seams or mines to achieve desired properties; and finally briquette required size for home heating.

Figure 3 is showing the standard deviation of heating value of coal, which were taken from major equipments in the typical briquette plant process in Figure 2. Finally, Figure 1 shows a relation between ash content and heating value.

Both domestic and imported coal are to be combined and fed to screen and the oversize from the vibrating screens (+10 mm) will be conveyed to crushers where it will be reduced in size and recirculated through the trommel screens to aforementioned vibrating screens, the underflow products of screens are taken to three different type of mixers to provide blending so that feed for each briquette is of uniform quality throughout the briquettes.

Table 1. Proximate analysis of feed coals at typical briquette plant

Type of coals	Heating value (Kcal/kg)	V.M. (Per cent)	C (Per cent)	ASH (Per cent)
A	4,640	4.67	56.48	38.85
B	4,070	5.83	48.71	45.48
C	4,170	5.78	49.97	44.25
D	2,700	6.14	32.41	61.45
E	5,490	5.82	65.30	28.88

* A.B.C.D.: Domestic coals in hoppers   V.M.: Volatile matter
      E.: Imported coals in hoppers   C.: Carbon

Table 2. Production capacity and distribution of briquette plants among the regions

Production (1,000/DAY)	Seoul	3 Major cities	Other cities	Total	% No. of plants	% Production
LESS THAN						
LESS THAN 50 (46,000 TON/Y)	–	1	126	127	48.7	5.5
50 – 100 (46,000 – 92,000 TON/Y)	–	–	45	45	17.2	4.8
100 – 200 (92,000 – 184,000 TON/Y)	–	–	33	33	12.6	6.4
200 – 300 (184,000 – 276,000 TON/Y)	2	4	9	15	5.8	6.7
300 – 500 (276,000 – 526,000 TON/Y)	7	8	6	21	8.0	27.4
500 – 700 (526,000 – 736,000 TON/Y)	1	4	6	11	4.2	19.2
OVER 700 (OVER 736,000 TON/Y)	7	–	2	9	3.5	29.8
TOTAL	17	17	227	261	100.0	100.0
% of NO. OF PLANTS	6.5	6.3	87.0	100.0	–	–
REGIONS PRODUCTION	41.4	21.8	36.8	100.0	–	–

Table 3. Standard deviation of heating value at coal piles

Unit : Kcal/Kg

Coal pile	Means (kcal/kg) (X)	Standard deviation (σ)	Range (R)	Remarks
Pile A	5,190	420	900	Domestic Coal
Pile B	4,760	70	530	Domestic Coal
Pile C	3,630	360	1,120	Domestic Coal
Pile D	5,520	290	980	Imported Coal

*   40 samples were taken from 4 coal piles "H" briquette plant in the Daejeon area.

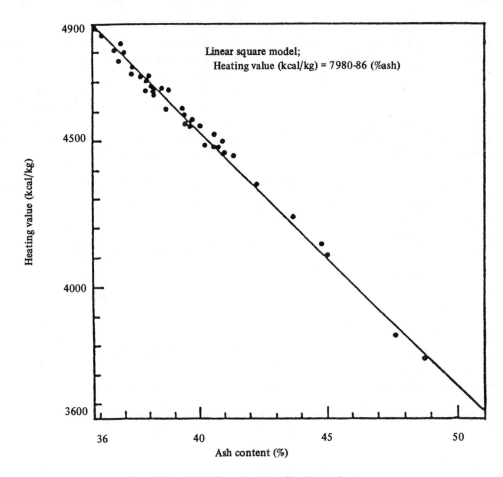

Figure 1. Ash-content vs. heating value.

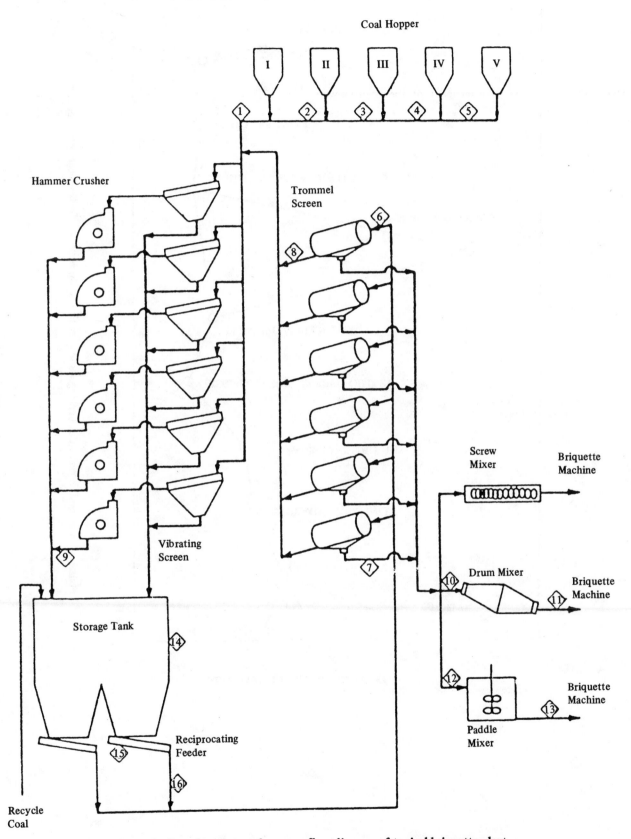

Figure 2. Sample points and process flow diagram of typical briquette plant.

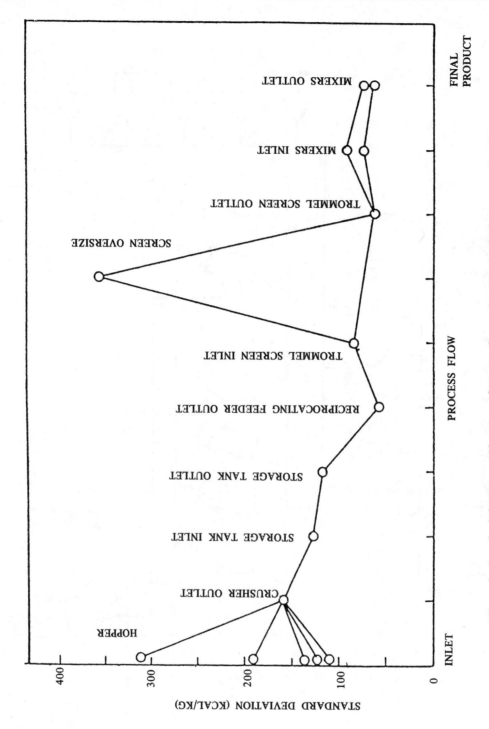

Figure 3. Standard deviation of heating value at major equipments in the briquette plant process.

3. Potential for Asian trans-country power exchange and development*
(E/ESCAP/NR.14/7 and corr.1)

Introduction

The development of interconnections between electric power systems of neighbouring countries is a worldwide phenomenon. It was started some time ago in other parts of the world, such as Western and Eastern Europe, North America and Scandinavia. In some Asian subregions of ESCAP, although limited exchanges of electric energy are taking place on a bilateral basis between a few neighbouring countries, no significant work has so far been done towards developing an integrated grid system. There is good potential for such a development in Asia.

Interconnection between power systems is an effective means of improving the overall level of system reliability and economy. Reliable and economic operation can be obtained through:

(a) Taking advantage of the demand diversity that exists in the system loads and outages of the component power systems, which will allow each national system to operate on less reserve than would normally be required;

(b) Co-ordinating different types of generation;

(c) Sharing power system risks;

(d) Optimizing capital investment;

(e) Having the possibility of adding more economic, larger units to a national system, maintaining a convenient reliability level at a smaller cost.

Recent changes in the energy scene have a considerable influence on the role and importance of regional interchange of electrical energy. The question of interconnection is linked to the development of large resources for power generation, such as hydropower, coal, natural gas and, possibly, nuclear electricity. To overcome such technical problems as transportation and handling or grid size, trading of these resources in the form of electricity is a viable option worth considering. However, such a development has to address several issues. The first and immediate need is the political will to act, followed by the need to address financial, institutional and technical problems.

The ESCAP secretariat, under its regional energy development programme, undertook a study covering just six countries of Asia to explore the possibility of interconnecting the electricity grids of these countries. The study, which was reviewed by an expert group meeting and a high-level meeting, brought out the possibility of such an interconnection. Drawing on the findings of the study, this paper analyses the prospects for expanding its scope to the whole of Asia and reviews experience gained worldwide in this field and the current situation in the region.

(a) Philosophy and definition

Because of the location of some electricity generating plants far from the demand centres, difficulties in storing electric energy, and the economic advantage of larger units, the interconnection technique was introduced to connect production units with consumers. This led first to the development of national grids. Trans-country power exchange is just an extension of this trend across the political boundary of a country. Trans-country power exchange networks are now in operation in many parts of the world. The totally integrated European, North American and Scandinavian electric power systems are good examples of the successful exchange or trade of electricity among a number of utilities.

Depending on the degree of involvement, there may be various types of co-operation among two or more utilities through interconnection of their systems to supplement each other for optimum economic operation. Electric power co-operation has various potential advantages: (a) optimum utilization of electricity-generating plants; (b) improved security of supply with smaller individual reserves; (c) diversity of resource options for power generation; and (d) mutual assistance during disturbances and emergency.

In a newly-interconnected system, the co-operation should be built up in various phases. Over time, as experience is gained, the degree of intensity could be increased.

(i) Power co-operation with common or unilateral sharing of spare capacity

Under such co-operation, the networks are interconnected and operated in parallel with the same frequency. In general, no power is exchanged except in the case of a disturbance. This energy is returned immediately after re-establishment of the equilibrium of the system. The networks share the spare spinning reserve as well as the stand-by reserve. All partners in the interconnection benefit from: (a) increasing the utilization period of the power stations; (b) reduced spare capacity; and (c) the possibility to co-ordinate the maintenance programmes of the power stations.

This type of interconnection requires agreement concerning the scope and conditions of energy supply and concerning technical and operational adaptation in areas such as frequency control, exchange of energy and mutual assistance. During this phase, the interconnection could even remain open.

(ii) Power exchange with short-term and long-term energy exchange

Such co-operation aims to balance, to a large extent, the different production and load characteristics of the

---

* Note by the ESCAP secretariat.

participating networks. The partners remain economically completely independent; however, co-operation is closer and the operations are dependent upon one another. The operation of power stations is mutually agreed upon according to the least cost.

(iii) Power exchange with transfer of power and common construction of overhead lines

In such co-operation, the participating partners agree to transfer electric energy through their networks for power exchange with the neighbouring networks, or they may jointly construct transmission lines. The partners still remain economically independent; however, concerning planning and investments they take into consideration the data of the interconnected system.

(iv) Power exchange with common construction of power stations

The independent partners agree to construct and to operate power stations together, or to erect power stations alternately. This form of co-operation allows construction of the largest technically and economically feasible unit and choice of an optimal site in the territory of one of the partners. This power station is operated either by a common pool sharing the profit of the station, or each partner uses his own dispatch centre and a common economic dispatch centre recommends the optimum use of the power station for each partner.

(b) Overview of experience worldwide

In 1984, over 80 per cent of the capacity of the world's electric power stations was concentrated in the fully-developed national power systems covering practically the entire inhabited territory of the United States of America, the Union of Soviet Socialist Republics, Japan, Canada and the European countries. Huge unified inter-State systems, the North American system, the West European system, the East European system, with capacities measured in hundreds of millions of kilowatts have been created.

By the end of 1985, the installed capacity of the electric power stations of all the Economic Commission for Europe (ECE) countries (excluding the USSR and the United States) was, according to ECE data, over 732 GW. Four groups of interconnected systems exist in Europe: (a) the interconnected power systems of the West European countries belonging to the various international associations and bodies for co-operation in the field of electric power; (b) the interconnected power systems of the Scandinavian countries; (c) the interconnected power systems of the Council for Mutual Economic Assistance (CMEA) countries, including the unified power system of the USSR; and (d) the power systems of the United Kingdom of Great Britain and Northern Ireland and its interconnection with France.

The figure shows the volume of electric power exchanges that took place in 1985 among various countries of the above groups.

The progress of system interconnection in Western Europe is co-ordinated by various associations and international bodies, including the International Union of Producers and Distributors of Electrical Energy (UNIPEDE), the Union for the Co-ordination of the Production and Transport of Electric Power (UCPTE), the Scandinavian Committee for Power Supply (NORDEL) and the Regional Group for Co-ordinating the Production and Transmission of Electricity between Austria, Greece, Italy and Yugoslavia (SUDEL). There are regular exchanges of power between the systems concerned.

In Eastern Europe, the power systems of the CMEA countries (Bulgaria, Czechoslovakia, the German Democratic Republic, Hungary, Poland, Romania and the unified system of the Southern USSR) work in parallel, in which respect the regime-planning and operational activities of the State dispatching offices of the various national systems are co-ordinated by the central dispatching office of the unified power systems. The power systems of the CMEA countries are developed under a single economic plan, and improvements are constantly being made in the structure and working conditions of the interconnected systems. Thus, the problems of the siting of new generating capacity and the construction of power lines are solved jointly.

Mutually-advantageous exchanges of electric power have been taking place for some time between the power systems of Eastern and Western Europe, although technically they are tied to the need to bypass isolated regions, which reduces the reliability of supplies.

Increasing the volume of and improving the conditions for electric power interchanges are of interest to all European countries, particularly in view of the world energy situation.

Recognizing the need for joint effort to solve fuel and energy problems, the European States established a good basis for co-operation in this field with the signing of the Final Act of the Conference on Security and Co-operation in Europe in 1975. Among projects of common interest States participating in that Conference gave priority to exchanges of electrical energy within Europe with a view to utilizing the capacity of the electrical power stations as rationally as possible. The participating States considered that possibilities existed in this field for projects of common interest with a view to long-term economic co-operation.

(c) Current regional situation

As pointed out earlier, limited exchanges of electrici y have been taking place bilaterally between neighbouring countries in certain areas of the Asian subregion.

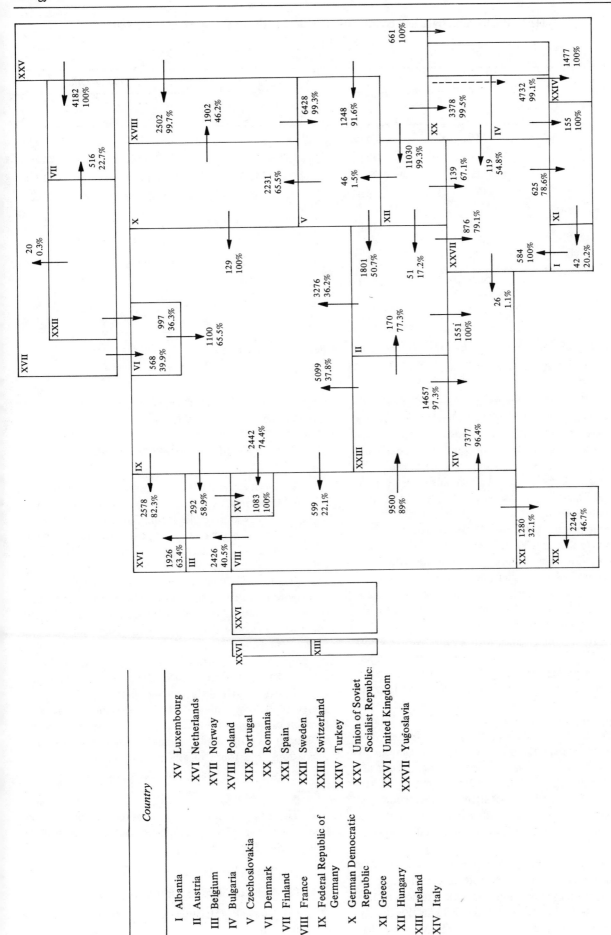

Figure. Annual balance of electric power exchanges between European countries in 1985 (GWh)

Source: The Electric Power Situation in the ECE Region in 1985. Document No. EP/R.123/Add.2, a report by the Economic Commission for Europe.

	Country		
I	Albania	XV	Luxembourg
II	Austria	XVI	Netherlands
III	Belgium	XVII	Norway
IV	Bulgaria	XVIII	Poland
V	Czechoslovakia	XIX	Portugal
VI	Denmark	XX	Romania
VII	Finland	XXI	Spain
VIII	France	XXII	Sweden
IX	Federal Republic of	XXIII	Switzerland
	Germany	XXIV	Turkey
X	German Democratic	XXV	Union of Soviet
	Republic		Socialist Republic:
XI	Greece	XXVI	United Kingdom
XII	Hungary	XXVII	Yugoslavia
XIII	Ireland		
XIV	Italy		

Table 1. Production, trade and consumption of electricity in the ESCAP region, 1984
(GWh)

Country or area	Production	Imports	Exports	Consumption Total	Consumption Per capita (kWh)
Bhutan	30.0	6.0[a]	–	36.0	26
China	376 990.0	250.0[a]	–	377 240.0	366
Hong Kong	17 923.0	–	740.0	17 183.0	3 125
India	165 440.0	3.0[a]	65.0[a]	165 378.0	221
Lao People's Democratic Republic	990.0	30.0[a]	690.0[a]	330.0	76
Malaysia	13 700.0	50.0[a]	0.0[a]	13 750.0	904
Mongolia	2 206.0	400.0[a]	–	2 606.0	1 408
Nepal	350.0	80.0[a]	6.0[a]	424.0	26
Singapore	9 401.0	–	73.0	9 328.0	3 672
Thailand	22 029.0	710.0	22.0	22 717.0	449

Source: 1984 Energy Statistics Yearbook (United Nations publication, Sales No. E/F.86.XVII.2).

[a] United Nations Statistical Office estimate.

Table 1 shows the countries and magnitude of electricity trade or exchange in the ESCAP region in 1984. These exchanges have been taking place mostly through weak ties in border areas of neighbouring countries. Experience in the region, though limited, is significant. The example of electricity trade between the Lao People's Democratic Republic and Thailand is a case in point: Thailand imports almost the entire energy of the 100 MW Nam Ngum hydropower station of the Lao People's Democratic Republic. An interconnection of 100 MW exists between Thailand and Malaysia. Because of oscillation problems, this connection does not allow parallel operation of the two networks. Solutions, including a DC back-to-back tie, are being studied. Since 22 December 1985, the 230 kV submarine cable link between Malaysia and Singapore has been in operation with 2 cables with a capacity of 2 x 250 MVA. The connection is at present used to provide additional spare generating capacity with zero energy exchange under normal conditions. In South Asia, Nepal imports power from India at 11 transfer points along the India-Nepal border and exports to India at a few points. Bhutan also imports a part of its electricity from India. The Interim Committee for Co-ordination of Investigations of the Lower Mekong Basin has been working to promote co-operative development of the hydro potential of the Mekong river to benefit its riparian member countries. Development of this large resource requires the collaboration of the countries concerned despite political differences.

In the ESCAP region, some progress has been made in the ASEAN countries to promote interconnection of their electricity networks. Some specific projects in this respect have been identified by the ASEAN Electricity Network Interconnection Group. An ESCAP study has further intensified the Group's activities by suggesting several projects for its consideration.

(d)   Study on trans-country power exchange and development

Under its regional energy development programme, the ESCAP secretariat undertook a study on trans-country power exchange and development. The main objectives of the study were to examine the possibility of subregional interconnection of electric power grids of neighbouring countries for the sharing or trading of electrical energy, and to explore the possibility of developing the large hydropower potential in those subregions. The study covered two subregions: subregion I, comprising Indonesia, Malaysia, Singapore and Thailand; and subregion II, comprising Bangladesh, India and Nepal. However, as India did not participate in the study, analyses in subregion II did not cover India in any detail. In subregion I, the interconnection is possible at only a few well-defined points, but in subregion II long common borders exist and therefore the location of possible trans-country interconnection depends on conditions other than geographic, such as the location of substations, the route of the existing high voltage lines, the load distribution, and the accessibility of the border region.

The study analysed five possibilities for interconnection technically and economically and two possibilities indicatively. These are:

(1) Thailand-Malaysia interconnection

(2) Malaysia-Singapore interconnection

(3) Sarawak (Malaysia)-Kalimantan (Indonesia) interconnection

(4) Sumatra (Indonesia)-Peninsular Malaysia interconnection

(5) Nepal-Bangladesh interconnection

(6) Nepal-India interconnection (indicative)

(7) India-Bangladesh interconnection (indicative)

The study also made an assessment of the hydropower potential of Nepal and of Sumatra and linked it with the interconnection projects for possible trade or sharing with other countries.

(i) Thailand-malaysia interconnection

The existing link between the two countries, with a nominal transformer capacity of 2 x 66.7 MVA with 115/132 kV, is weak for an effective power exchange. After careful analysis, a transmission capacity of approximately 300 MVA was favoured in the study. The economic evaluation compared the present value of investment for the interconnection with that of an equivalent power station and it appeared that a high internal rate of return could be obtained with the interconnection. Two cases of operation were studied, one with the assumption that the link serves only to provide spare capacity and the other with the assumption that energy is exchanged. The study demonstrated that favourable conditions for power exchange exist because of the different patterns of the daily load curve in the two countries. This situation would allow exchange of power on a daily basis with a reduction of the annual peak by at least 6 per cent. The conditions for a daily power exchange are particularly favourable if the peak loads of the National Electricity Board (NEB) of Malaysia and the Public Utilities Board (PUB) of Singapore are balanced with those of the Electricity Generating Authority of Thailand (EGAT).

(ii) Malaysia-Singapore interconnection

The interconnection between the two countries has been in operation since December 1985 with two cables of 230 kV, each equipped with an auto-transformer of 250 MVA. The cable link is operated with zero power exchange, i.e. spare spinning and stand-by reserve are shared and unavoidable energy exchange was returned immediately. A power exchange between the networks of NEB and PUB is possible if surplus energy available in one network is produced at a lower cost than the energy generated in the other country.

(iii) Sarawak-Kalimantan interconnection

Detailed technical studies have been elaborated by the Joint Task Force of the National Electricity Enterprise (PLN) of Indonesia and the Sarawak Electricity Supply Corporation (SESCO) of Malaysia to define and evaluate the interconnection of the network of SESCO with that of PLN after the completion of the Bakun hydropower project in 1995. The economic evaluation included in the ESCAP study calculated the economic benefit of SESCO and PLN operating the interconnection even before the completion of the Bakun project, or, in the event of any possible delay in the project, transferring excess power from Batan Ai power station, sharing the spare capacity of both networks and exchanging power on the basis of the different patterns of the daily load curves.

(iv) Sumatra-Peninsular Malaysia interconnection

For the exchange of electricity with Malaysia, it was proposed to develop the hydropower potential of the Asahan River with a total installed capacity of 592 MW and 3,364 GWh per year of firm and secondary energy, and other sources of energy, such as coal. Two alternative transmission-line routes to interconnect the networks of PLN and NEB were studied. For the first alternative, only 50 km of submarine cable would be needed, while the other reguires 173 km of submarine cable.

(v) Nepal-Bangladesh interconnection

Even in the distant future, Nepal will be able to utilize only a small fraction of its very large hydropower potential (over 83,000 MW). Thus a large quantum could be used for export. For export to Bangladesh, the potential of the Kosi River basin and the Arun River in particular might be developed. However, the interconnection between Nepal and Bangladesh can only be realized with permission to cross Indian territory with a transmission line.

(e) Observation and recommendations by the Expert Group Meeting and the High-Level Meeting on Trans-country Power Exchange and Development

The study on trans-country power exchange and development was reviewed by the Expert Group Meeting on Trans-country Power Exchange and Development, held in September 1986, which made several observations and recommendations for consideration by the High-level Meeting on Trans-country Power Exchange and Development, held in November 1986. The Expert Group Meeting in general commended the secretariat's draft study and appreciated the efforts of ESCAP/REDP (regional energy development programme) to foster closer co-operation amongst utilities in the Asian region. Some specific comments were made to improve the quality of the report. Those comments were taken into consideration in the preparation of the final report which was published as a United Nations publication[1] in May 1987.

The Meeting in general was in agreement with the findings of the study. However, it recognized the need for independence of individual utilities with respect to their own development and operation. It considered the study as a basis for further in-depth studies and/or follow-up action. It was suggested that issues such as organizational aspects,

---

[1] ST/ESCAP/474.

trade and tariffs should be treated in more depth in any further study. Regarding the tariff structure, however, the Meeting noted that tariffs were a matter of agreement between two partners and should therefore be tailored to their mutual benefit.

The Meeting reviewed the report project by project and offered its comments. For the Thailand-Malaysia interconnection project, the Meeting discussed the technical feasibility of a DC alternative, not chosen in the study, and agreed that it should not be eliminated before detailed studies had been made. The Meeting was informed that some of the recommendations of the study were already under consideration by the two countries involved. Concerning the Malaysia-Singapore interconnection, the Meeting was informed that the existing interconnection was performing well and the experience gained might be of interest to others. For the Sumatra-Peninsular Malaysia interconnection, the Meeting was informed that other energy resources, such as coal, could possibly be developed as an alternative means of power generation for export. Regarding the Nepal-Bangladesh interconnection, the Meeting noted the interconnection could be realized only with permission to cross Indian territory with a transmission line.

The High-level Meeting considered the report of the Expert Group Meeting. The Meeting recommended that further studies should be undertaken in the second cycle of the regional energy development programme. Other external assistance should also be made available. The High-level Meeting made general and project-specific recommendations and observations concerning both power exchange and the development of resources for power generation. These recommendations are meant for regional and international organizations as well as for the countries concerned. Highlights of some of the recommendations are noted below:

(a) As a follow-up to the study, the long-term prospects for the utilization of other sources of energy, such as coal and natural gas, for electricity generation should be studied in the second phase;

(b) Other interested countries, not covered in the study, should be included in future studies;

(c) To promote exchange and dissemination of information and experience on energy exchange and systems interconnections between subregions or systems, activities such as seminars and study tours should be organized;

(d) A study should be carried out to determine the optimum pool size of the interconnected systems and the timing of each stage of the interconnection project, as well as the possibility of setting up a co-ordinating centre for load-dispatching in the region;

(e) The implementation programme should take into account various technical, economical and political factors, including each country's economic and political independence;

(f) In taking note of the kind of assistance available from the Asian Development Bank, the World Bank and other financing institutions, the Meeting recommended that concerned Governments undertake several follow-up studies under bilateral or multilateral arrangements.

(f) Prospects for an Asian grid

Should the countries be interested, an Asian integrated electric power system like those in operation in Europe and America is possible. It could bring benefits for all, and would foster concrete co-operation among the countries. There are enough energy resources region-wide that can be developed and traded through the Asian integrated electricity system network if the countries co-operate and participate actively. The first and immediate requirement is political will. As reviewed in earlier chapters, some initiatives have already been implemented among a few countries, particularly ASEAN countries. Some impetus is necessary to enhance these initiatives. Any project involves bilateral agreements, but prospects for stronger and more economic interconnections depend on the co-operation of a group of countries initially at the planning of interconnections, followed by joint studies on profitability and the technical possibilities. Finally, co-ordinated operation of the integrated system is also important for the economy of the region. The extent and coverage of this operation will depend upon such factors as techno-economics, reliability and security of supply. It should be noted, however, that such developments take time even when different provincial utilities in the same country are involved; this is even more the case with sovereign States.

(i) Extent of involvement

As briefly explained in section I, philosophy and definition, the degree of interconnection can vary from zero energy exchange up to power exchange with common construction of power stations. The prospects for collaboration in the Asian subregion fall into two groups. There can be collaboration on specific projects between two or more partners and collaboration among most, if not all, countries and territories in Asia. An outstanding and mutually-advantageous example of collaboration could be the development of large hydropower resources. The hydropower potential of the Mekong River will far exceed the combined maximum demand of its riparian countries for a long time to come. The hydropower potential of Nepal cannot be optimally developed and utilized without securing a power market in the neighbouring countries. The hydropower potential of Sumatra and Bukun (Sarawak, Malaysia) is considerable and could be developed and shared through a prospective ASEAN grid. The Salaween River in Burma has a large potential, the development of which requires collaboration with Thailand, and possibly with the members of the proposed ASEAN grid. There are also good prospects of exploiting other sources of energy, such as coal, natural gas or even nuclear energy, for electricity generation for common use. In the Expert Group and High-level Meeting on Trans-country Power Exchange and Development participants in fact recommended that the potential of such developments should be explored. Coal and natural gas, the transport and handling of which are sometimes difficult and need additional infrastructure, could be used to produce electricity at the mine or field, which could be exported to load centres via an interconnected grid. In this way a good mix of hydro-thermal system could be built for optimizing the benefit of all participating countries. Larger, highly-efficient, power plants could also be accommodated in an integrated system. In the longer term, the nuclear energy option could also be considered for the common benefit.

Optimum utilization of resources and power systems would involve stronger interconnection between countries. Thus, although initial involvement may be low-key, step-by-step progress towards a stronger grid could bring good results. The efforts being made by the ASEAN Network Interconnection Group are noteworthy.

Financing agencies such as the Asian Development Bank, the World Bank, and the Canadian International Development Agency (which participated actively in the first ESCAP study) have strongly supported the idea of interconnection and joint development of resources, if the countries concerned are interested. Thus one of the major constraints, financing, may not pose a serious problem.

(ii) Coverage of the grid

As mentioned earlier, one of the salient recommendations of the Expert Group and the High-level Meeting is that other interested countries not covered in the study should be included in future studies. This implies that there is a general interest in a greater area coverage. The ESCAP study covered only six countries, four in South-East Asia and two in South Asia. In South-East Asia, specific recommendations have already been made to include Brunei Darussalam and the Lao People's Democratic Republic in subregion I (Indonesia, Malaysia, Singapore, Thailand) of the study. The active participation of India (which did not take part in the first study), Pakistan, Afghanistan and the Islamic Republic of Iran in subregion II of that study (Nepal, Bangladesh, India) could lead to a South-Asian grid. Burma, being in the middle, might join either subregion, or the ASEAN grid might be connected to the South-Asian grid through Burma. In the longer term the Philippines intends to connect its system with Malaysia. Through the Mekong project, member countries might also connect their systems with the South-East Asian grid. Although a good deal of work has to be done to exploit such a possibility, exploratory efforts should be started as soon as possible. The High-level Meeting drew up a tentative schedule of follow-up activities in trans-country power exchange as shown in the annex. This may look ambitious but, with the concerted efforts of all concerned, is not impossible.

(iii) Technology available

As mentioned earlier in this paper, interconnected systems have been in operation in a large number of countries and for quite some time.

Over the years the technology has advanced with the progress of the electric power systems. These advances have made interconnection even more feasible. Extra high voltage (EHV), ultra high voltage (UHV) and high voltage direct current (HVDC) technology has made possible the transmission of larger amounts of energy over increasingly long distances with relatively low losses. Advances in the area of submarine cables have made it possible to connect two systems separated by waterways and a number of other technical problems can now be solved through advanced technology.

(iv) Limitations to consider

The building of a new interconnection causes a change in the systems connected. It will have considerable impact on the original systems, both technically and economically. In order to provide good solutions, detailed studies have to be made to examine various factors and limitations, including the impact of the technical behaviour of the power systems and influence on other parameters such as reserves or reliability. Technical characteristics of the systems that may change drastically in some cases include: (a) transient and dynamic behaviour; (b) frequency control and spinning reserves; (c) transient stability and control; (d) reliability. They were discussed at the recent Asian Energy/Power Technical Seminar, Equipment Exhibition, and Optimization of Electric Power Systems Workshop (Kuala Lumpur, 8-11 April 1987).

(g)   Summary and conclusions

(1) The techno-economic advantages of interconnected systems are enormous. A good integrated network means savings in power-generation investment and operating costs, and thus cheaper power.

(2) Interconnection may begin with minimal involvement of participating utilities, gradually rising to stronger integrated ties.

(3) Interconnection between power systems is an effective means of improving the overall level of system reliability and economy.

(4) Interconnected systems are in successful operation in North America, Western Europe, and Eastern Europe.

(5) Currently, limited exchanges of electricity are taking place between a few countries of the Asian subregion.

The ASEAN Electricity Network Interconnection Group has been working actively to initiate the ASEAN grid.

(6) An ESCAP study on trans-country power exchange and development made a preliminary assessment of the possibility of subregional interconnections of electric power grids in selected countries of Asia. The result of the study indicates positive prospects for interconnection in the countries studied.

(7) The Expert Group Meeting and the High-level Meeting on Transcountry Power Exchange and Development reviewed the study and agreed with the findings of the study. Recognizing the potential of interconnection, both the meetings formulated a number of recommendations and suggestions for follow-up action to promote Asian interconnection.

(8) From all indications, it may be concluded that the prospects for Asian trans-country power exchange are bright. However, such a development is not possible unless a number of issues are resolved. Besides the political will to act, some financial and technical problems also have to be addressed. Co-operation between participating countries is very important during both planning and operation.

(9) In the current situation, the national and provincial electric power utilities of the Asian countries are developing fast and this is the right time to incorporate any interconnection possibilities into their system planning. To draw on the techno-economic advantages of interconnection, it is desirable that countries and utilities start co-operation among themselves right from the planning stage. Joint studies have to be made on the profitability and technical possibilities of interconnections.

(10) It is advisable that interested countries and power utilities have a small but flexible common organization or group that can serve for direct contacts between planners at various levels. An open-ended TCDC working group of national and provincial electric-power utilities might serve such a purpose. The ESCAP secretariat stands ready to help organize such a working group.

Annex

TENTATIVE SCHEDULE OF FOLLOW-UP ACTIVITIES
IN TRANS-COUNTRY POWER EXCHANGE

ESCAP	*Subregion I countries (ASEAN)*	*Subregion II countries (South Asia)*	
1. April 8 to 11, 1987 Kuala Lumpur: Optimization of Electric Power Systems Workshop	Specific feasibility study for firm interconnection of EGAT (Electricity Generating Authority of Thailand) with the NEB (National Electricity Board of Malaysia)/PUB (Public Utilities Board of Singapore) network (possibly for Asian Development Bank funding)	Bilateral discussion India/Nepal, Nepal/Bangladesh, Trilateral discussions India/Nepal/Bangladesh to agree on approach	*1987* *1988*
2. Electric futures study under REDP, cycle II	Specific longer-term pre-feasibility studies in (a) Kalimantan/ Sarawak, (b) Sumatra/Malaysia, to include effects of other than hydro resources	Study of Arun 2 (feasibility) Nepal, India and Bangladesh optimization of electric grids (national studies)	*1989* *1991*
	Longer-term study for the Philippines (possibly for bilateral funding)	Possible large hydropower projects	
	Full interconnection, Association of South-East Asian Nations	Full interconnection, South Asia	*2005*

4. Human resources development issues*
(E/ESCAP/NR.14/8)

(a) Manpower needs of the energy sector

The critical nature of the energy problem in rural areas of the developing countries is viewed today with great concern and increasing urgency. The United Nations system, together with national Governments, has expended a significant amount of effort to cope with energy problems, particularly in the rural areas. Consequently, substantial programmes and projects have been drawn up to accelerate the application of new and renewable sources of energy for a sustainable energy future.

Many constraints, however, inhibit the achievement of the desired results. Among these, the lack of adequately trained manpower for effectively implementing national energy development programmes and projects is a constraint of key significance in energy sector programming. This manpower gap affects, although with different degrees of severity, all categories at all levels. The human resources bottle-necks of energy development are more obvious in the implementation of new and renewable sources of energy projects related to new technologies, in which in some cases no more than 10 per cent of capital completion in the five years 1981-1986 had been achieved.

The size of an adequate energy training effort

There is now an urgent necessity for most developing countries in Asia and the Pacific to intensify their efforts to increase energy production from indigenous sources. This will require substantially more manpower, with suitable skills and qualifications to plan, develop and carry out energy projects while efficiently organizing and managing the overall energy sector.

Some indication of how important it will be to develop human resources can be seen in the potentially rapid growth of investments required in the energy field over the coming years. According to the Nairobi Programme of Action for the Development and Utilization of New and Renewable Sources of Energy, such sources of energy will account for a substantial and growing proportion of these investment needs. Developing countries spent a total of around 34 billion dollars for commercial energy production and transformation in 1980, nearly three times the average expenditure for the period 1966-1975. The principal investment requirements in commercial energy for the period 1981-1990 are expected to quadruple to about 136 billion dollars. According to an ILO study, the investment needs for supporting action, including feasibility studies in the field of new and renewable sources of energy, are estimated at 14.2 billion dollars for the period 1982-1990. Regardless of the grave uncertainties about external capital,

---

*   Note by the ESCAP secretariat.

and even if targets are not fully realized, it is clear enough that the success of such a massive energy effort will depend to a considerable degree on the mobilization of human resources.

(b) Identification of areas where energy training activities are urgently needed

Energy sector development activities are diverse; there is a whole range of activities related to obtaining energy materials and tapping potential new fuel, as well as activities in the use of energy.

Human resources development to meet the requirements of the energy sector needs to identify with operationally meaningful precision where or in what specific components of the energy sector development programme manpower shortages exist and are likely to develop; whether the existing training facilities are turning out the requisite skill groups in sufficient numbers and at the required time; and what specific manpower/labour market policies might help in overcoming imbalances in supply and demand of various skill categories, now and in the future. It is therefore necessary to make realistic assessments of manpower requirements in the energy sector. Such a manpower and training needs assessment activity would need to diffuse fairly precisely the coverage that is intended and possible.

The coverage of manpower assessment should be decided in three directions: by technologies, by economic activities and by manpower categories.

Technologies differ largely in the energy resources used. They differ widely in capital and working costs, operational techniques, manpower requirements, in qualitative and quantitative terms and by occupational mix. The choices of energy technologies available are: (a) conventional and (b) non-conventional. (Such categorization, although always to a certain degree arbitrary, reflects current United Nations usage.)

Conventional categories are those related to coal, oil, natural gas, nuclear electricity and hydropower development.

Non-conventional or new and renewable source categories are those related to solar energy, geothermal energy, wind power, tidal power, ocean wave power, ocean thermo-electricity, biomass conversion, biogas, dendrothermal power, etc.

The technology also varies in almost every case depending on whether direct application into heat, light and mechanical power is desired, or electricity is required through the process of conversion. Thus trained technical personnel are necessary in sufficient numbers, and with the required capabilities.

The core group of economic activities may include prospecting and assessment of energy resources; actual extraction of energy materials (operations of drilling and mining); intermediate processing (coal washing and grading, oil refining, gas separation, materials concentration, pelletization, etc.); conversion to electricity (electricity generation), transformation, power transmission and distribution.

The linkage activities may include survey, appraisal and design of facilities; construction of energy facilities; storage, transportation and containerization of fuels for end-use; consultancy and training on feasibility studies, management of capital work projects and operational energy establishments, and energy conservation at enterprise levels; manufacture of energy-related equipment; research and development; planning for the energy sector, energy resources management and conservation policies and programme development; training and manpower development services.

The main groups of manpower in the energy/power sector, like any other sector, are: (a) managerial, professional, and highly educated and experienced manpower; (b) technical and supervisory manpower; (c) skilled manpower; and (d) semi-skilled manpower.

It is therefore necessary first to assess the needs of different types of manpower in a particular level of energy technology before formulating a manpower development plan. It is often found that a large number of personnel are required in the last three groups, but emphasis is often disproportionately placed on the first.

(i) Energy planning manpower

Energy sector planning and management call for imaginative and creative thinking, ideally by teams of local and foreign specialists able and willing to devote sufficient time and effort to an in-depth understanding of the economic and social context before applying their own experience and competence to the solution of the problem at hand. Unfortunately, the danger of superficial assessments or appraisals seems to be growing as the increasing workload of most international agencies active in this field leads them to assign energy assessment work to relatively junior staff with the highest academic training but little or no field or management experience. As their contacts in the country under study are all too frequently limited to a small number of civil servants, also with a primarily academic outlook, the risk exists of an "empty dialogue" somewhat disconnected from the realities of energy management, a particularly serious shortcoming in the case of renewable energy programmes requiring the co-operation of village communities.

Such an "empty dialogue" shortcoming could somehow be alleviated through in-country training activities to produce village-level energy researchers. The researcher is a catalyst for community self-study, question-raising and information-sharing. Through dialogue and involvement with all village groups, the researcher is a facilitator of negotiation and decision-making, who assists village groups in linking external technology and related support to local resource mobilization. Such researchers, who play a crucial role in the training programmes in the villages, should be given scientific training, and governmental and financial back-up. Exchange of project personnel among countries or providing the facilitators with an opportunity to participate in missions to other countries constitutes an approach to enhance the competence of the facilitators. Such an approach could reduce the reliance on external workers in some countries which are in need of manpower.

This human side of energy planning has other facets, one of which is of paramount importance. Training requirements for energy-sector managers at all levels are frequently underestimated, for both the sector as a whole and the various supply subsectors. This is largely due to the difficulty of estimating the needs *a priori*, in time for the newly-trained staff to become operational when the need arises. The most adequate mix of academic and practical training, in the country and abroad, can only be assessed in a very approximate measure, often based on the experience of other countries where problems are not totally identical. Moreover, the training component of foreign technical assistance loans or grants for energy efficiency or conservation programmes is all too often insufficient to cover total needs. A serious difficulty frequently encountered is that, once recruited and trained for a public sector energy entity, junior staff may be tempted to leave for better-paid positions in private industry. While this may not be undesirable from the viewpoint of overall efficiency of the economy, it is a frustrating and costly process for the agency, making the initial "guesstimate" of training needs even more difficult.

The need for a minimum of stability in the staffing of newly-created energy entities raises a related topic: the "identity" of energy planners. The term is vague enough to accommodate almost anyone's concept, from the lowest to the highest professional categories. An international agency puts energy planners in the lowest professional bracket, while another, based in a different part of the world, places them at the very highest level, considering that only a few persons in any given country qualify for the title. Much depends on whether or not planners are also policy makers, or even decision makers, rather than staff members of the programming department of a ministry or of an energy supply entity such as the local electricity authority.

(ii) Technology-specific manpower

Manpower preparedness for accommodating structural changes in the production sectors to adopt new and advanced energy/power technologies is just as important as other issues. Diversification in the use of energy sources in power plants and industries, and technological advancements such as new techniques and computer applications, are some of the areas where evolving changes are taking place. Inter-fuel substitution of oil for other alternative sources of energy, such as coal, natural gas, nuclear power, and new and renewable sources of energy will need structural adjustments of the whole energy sector, and the power sector in particular. The task is huge, and so are the manpower requirements. In addition to these, there are several constraints and limitations, such as large investment requirements, lack of adequate infrastructure, and much lower capabilities than are required.

Integrated energy planning, together with human resources development planning, is required. Long- and short-term human resources development plans (including training of trainers and on-the-job training) should be enhanced.

Energy sector manpower needs in the least developed countries of the region and in the island countries, in terms of the number of personnel and the expertise, are more acute than in other developing countries of the region. This is particularly true in the level of energy development and that of technology. Many of the needs are in basic fields such as energy resource assessment, project identification, and feasibility studies. The availability of energy experts is quite low, and many countries, particularly the Pacific island countries, depend heavily on expatriate specialists. Thus, training is needed in almost all levels of skills, starting from technicians and skilled workers to engineers and energy experts. Training is also needed at both the managerial and the working levels.

Rural electrification through utilizing local resources is becoming more economical in many cases, particularly for isolated areas. But sophisticated technologies such as solar photovoltaic application to lighting, refrigeration, social uses, solar-pumps, wind electricity generation, mini- and micro-hydropower, all need specially trained personnel. However, for the success of technological applications, mass participation is needed from among the users, in this case the rural people, and women in particular. The training needs are thus different. Training is needed to adapt technologies to rural conditions in their socio-economic environment. Here social acceptance of a particular technology is more important than other considerations. Thus awareness and motivation are key points in training. Other important points that should be kept in mind are the economic conditions and the general level of education in rural areas.

Following the recommendations made at the Nairobi Conference (1981) and the Regional Expert Group Meeting on New and Renewable Sources of Energy (1982), ESCAP has carried out many projects promoting co-operative research, development and demonstration on new and renewable sources of energy in the region through activities implemented by the Energy Resources Section and the regional network on biomass, solar and wind energy as well as the UNDP-funded regional energy development programme (REDP) and the Pacific energy development programme (PEDP).

Emphasis has been given to demonstration of newly-established facilities dealing with biomass and solar energy and small hydropower.

Various types of small and mini hydropower generation systems were demonstrated under the co-operation among the participating countries in the Regional Network for Small Hydro Power (RNSHP), in China, for promoting deeper awareness of this technology and highlighting necessary measures for application in each country. This led to the cross-country co-operation in setting up small hydro facilities and TCDC operation, such as co-operative technology transfer between China and Fiji.

In the field of solar energy application, there have been many co-operative activities in research, development and demonstration, such as solar drying in Indonesia and Thailand and solar thermal in Thailand, to name just a few. These activities resulted in the improvement of each country's facilities and upgrading their technical capability. Another example is in the area of solar photovoltaic (PV) electricity generation in which TCDC exchange of PV systems for evaluation is under way following up a recommendation of the Regional Expert Seminar on Solar Photovoltaic Technology held at Bangkok in June 1985. Countries advanced in this technology, such as China and India, would provide information on these systems, leading to TCDC co-operative programmes with interested countries in the region such as Indonesia, Maldivers, the Philippines, Thailand and Viet Nam.

Another co-operative research, development and demonstration project in solar PV technologies is under way based on the Tokyo Programme on Technology for Development in Asia and the Pacific which was adopted as one of the resolutions by the Commission at its fortieth session. This project consists of two phases of a demonstration-cum-seminar/training course organized in Indonesia and Pakistan. Following an expert group meeting in June 1986, the first programme started in Yogjakarta in mid-January 1987 with a seminar/training course on the evaluation, design and implementation of solar PV systems in developing countries. A similar project will be conducted in Pakistan in the first phase, and will be followed by more specialized courses in the second phase in 1988. A component for the Pacific island countries and Maldives, covering roving PV installers' training courses, has been added to the project.

As for solar thermal technology, workshops/training courses have been organized aimed at promoting regional co-operation. One of the typical examples was a training course on solar hot water systems jointly organized with the Asian Institute of Technology (AIT) in 1985. It was agreed in the course that co-operation should initially be focused on two areas: (i) adaptation of standards to suit regional conditions for solar collector testing and (ii) performance evaluation of solar systems. Following up on these, ESCAP is planning to conduct study missions to establish regional co-operation between countries in this field.

Reviewing the above activities undertaken by ESCAP, it is considered that regional co-operation in research, development and demonstration on new and renewable sources of energy has been successfully promoted in some cases. However, much more needs to be done. In order to accelerate regional co-operation, follow-up action should be taken in each field of new and renewable sources of energy. For instance, it is envisaged to be essential for providing necessary data to facilitate further application of solar PV technologies to conduct co-operative climatic measurement of solar insolation and its local characteristics, and to establish the standard testing method of efficiency for solar heat exchanging systems under regional co-operation, so that each country can improve the solar thermal facilities based on a common standard.

As for the field of research and development, there is still some room for promoting co-operation among the countries of the region, especially multilateral regional co-operation, in addition to the present ongoing bilateral co-operation. In order to facilitate the smoother transfer of new and renewable sources of energy technologies and their adaptations, it is necessary for the developing countries of the region to improve their technology levels so that they can catch up with developed countries where new technologies have mainly been developed. Research and development co-operatively conducted among the developing countries should foster improvement of their technological capability.

(iii) Energy management and conservation manpower

Many countries are attempting to diversify their energy mix through a policy of promotion of indigenous energy resource development and reducing the dependency on imported oil. Efforts in this matter should be supported by the formulation of an energy demand management strategy.

As an initial step, in some countries a continuous programme of establishing data base on sectoral energy demand and energy consumption patterns has been undertaken. Much work has been done in this matter using domestic resources, and the results could be used as input into the formulation of an energy demand management project if some methodological differences among the different surveys could be reconciled into a uniform and consistent approach.

For the next step it is planned to develop energy management activities in these countries and formulate an energy demand management strategy for the next 15 years to promote more efficient energy utilization combined with indigenous energy resources development. Many Governments are seeking expatriate technical and financial assistance for such undertakings. Advisory missions by ESCAP could continue to render such assistance. However,

other agencies may also be usefully involved (Asian Development Bank (ADB), World Bank), so as to expand the scope of the work, commensurate with the needs identified.

There is a clear need for more research on the comparative degree of effectiveness of each of the four types of public policy tools for demand management: regulation, incentives, information and training under varying economic and social conditions. With regard to incentives and disincentives, a study of the effects of taxation and pricing on energy consumption patterns in a selected number of countries of the ESCAP region, for example, would be very useful, possibly in connection with or as a follow-up of the recent survey by ADB of energy demand management in the region.

ESCAP itself could take the initiative of organizing a seminar or other form of meeting on taxation and energy management, or more specifically on the applicability and comparative effectiveness of a selected tax. One such tax which has proved particularly effective in curtailing industrial energy use in France, for instance, is the differential tax on industrial fuel oil, which could usefully be introduced in some countries wherever the private industrial sector is a sizeable part of the national economy. Other forms of incentives and disincentives to energy conservation successfully applied in Europe should be better known by planners in the ESCAP region, where little has been done so far to inform them on the subject.

Still another type of activity worth considering would be, as envisaged earlier, a study of issues connected with the creation of joint power markets by neighbouring countries.

This type of information and problem identification work would appear particularly fitting for ESCAP, upstream of more intensive and project-oriented studies supported or undertaken by international lending institutions such as ADB and the World Bank.

(iv) Statistics manpower

Energy statistics play a vital role in energy planning, policy-making, monitoring of energy resources and their optimum utilization. Systematizing the collection and dissemination of energy information is regarded by ESCAP as of the utmost importance. At the Asian Forum on Energy Policy organized by ESCAP in October 1986, the regional statistics adviser presented a paper identifying areas of general concern in the organization of energy statistics, and their collection, compilation and publication in Asian countries, based on the experience of seven countries that had sought his advice. The paper stated that energy statistics collection required a more detailed technical knowledge of the subject-matter than most, if not all, other statistics normally collected by Governments. Without initiation and considerable training, it was unlikely that those in a central statistics organization could meet the requirements of those in energy (and individual fuel) ministries. Conversely, those in energy (and individual fuel) ministries might be deficient in statistical theory and training, without which there might be loss of data precision, and speed and comprehensiveness of analysis; perhaps most important, the cost of statistics collection might be higher than necessary.

In order to obtain good energy statistics, and the greatest benefit from them, a Government needs to adopt the following:

(a) To make one centre (whether in a central statistical office or energy ministry) responsible for energy statistics collection, assembly and dissemination;

(b) To provide appropriate training in both energy and statistics matters for those directly involved in the preparation of energy statistics;

(c) To creat appropriate channels of communication within and between energy and individual fuel ministries to ensure that users' needs are being reflected in the statistics prepared;

(d) To eliminate not only any duplication of energy statistics collection, but also duplication of non-energy statistics of direct applicability to energy ministries (for example, in the industrial, transport and socio-economic statistics areas): this required good communication between energy and non-energy ministries;

(e) To establish confidence between the suppliers of energy data and those responsible for collecting it (through the delineation of the purposes for which the data may be used, and so on).

(c) Types of training requirements for further human resources development and implications for the training activities

A logical approach to practical implementation of training programmes lies in need assessment and identifying issues relevant to needs. Types and methods of training depend on the requirements. In the broader context, it may be the responsibility of the educational system to impart basic knowledge, through regular school and university curricula. That is a part of the national development plan. Utilities, however, are more concerned about the people already in their jobs, or those who are newly recruited. Post-entry level training may be short (or sometimes long if it involves higher education). Longer training is normally given by universities in regular programmes, or through special arrangements.

There are several ways in which short-term training can be given, among them on-the-job training and special training programmes. Most of these types of training are given in the utilities' own facilities, but can also be arranged through co-operation with other institutions in foreign countries. International organizations and agencies do offer such training opportunities. At senior levels, brief seminars, workshops and conferences also help to attain some train-

ing objectives. Such national events play a significant role in the quick dissemination of technical information. International seminars, workshops and conferences offer an excellent opportunity for sharing experience and information among participating countries and leading experts. Participating experts in turn can multiply the knowledge they gain through organized national programmes. Because of higher needs compared with resources available, and considering effectiveness, international organizations implement such courses frequently. A roving training course is another possibility, in which a group of experts conduct training courses country by country. The ESCAP secretariat has been organizing various types of seminars, training courses and workshops on a regular basis, covering all aspects of energy, such as power systems, oil and gas, coal, solar photovoltaics, and so on.

The need for human resources development in the energy sector is enormous but the resources available are limited. Thus, pooling of resources and exercising joint efforts to address this difficult issue offers one suitable solution to the problem of energy manpower development. Technical co-operation among developing countries and regional/subregional co-operation in energy manpower training may play a significant role in supplementing national efforts. Several such activities have already taken place and some others are planned. A few activities of the ESCAP-executed, UNDP-funded projects, i.e. the regional energy development programme (REDP) and the Pacific energy development programme (PEDP), have been endeavouring to foster intercountry co-operation. Another approach which is currently being pursued is the networking approach which appears to be cost-effective for information dissemination, sharing of expertise and the latest know-how via specific studies and training. The regional network on biomass, solar and wind energy in the ESCAP secretariat and the regional network on small hydropower in Hangzhou, China, are two examples of such networks being developed to cater for such needs. The Asian and Pacific Development Centre has also launched another network, the Asian and Pacific network on energy planning (APENPLAN). Recently, at the World Bank Seminar on Training of Trainers – Energy/Power Sector, held at Bangkok, the establishment of a network on the energy/power sector training was also proposed. UNESCO has the Co-operative International Network on Training and Research in Energy Planning (CINTREP).

There appears to be a great potential for regional/subregional co-operation in energy planning and development, including training, in the ESCAP region, achieving cross-fertilization of programmes and economies through the common use of existing facilities.

In March 1986, CINTREP organized the Workshop on Energy Conservation in Industry at Chulalongkorn University in Bangkok, Thailand. The survey of energy conservation measures in seven countries of the region presented at the Workshop, as well as the overview papers, resulted in an appreciation of the importance of further work in this area. In the light of this, both a policy-oriented workship and a technology-oriented workshop have been proposed to UNESCO as follow-up activities for the next biennium. While interest was expressed by the Republic of Korea in hosting the second of these workshops, ESCAP has undertaken to arrange the organization of the first one with the collaboration of Chulalongkorn University.

In the energy work programme of ESCAP for 1986-1987, a workshop on energy conservation policy and measures for energy demand management has been scheduled, with a publication entitled "Energy conservation policy and measures for energy demand management" to consolidate results, by the fourth quarter of 1987. The workshop is tentatively scheduled for October 1987, and the modalities of co-operation have been discussed with Chulalongkorn University and CINTREP, in addition to the German Agency for Technical Co-operation (GTZ).

Limited funds and a "synthesis" consultant are available from GTZ, and the ESCAP secretariat will support the workshop with professional resources, organization, technical editing of the subsequent publication, and so on.

(d)   The role of regional/national institutes linked in human resources development (exchange of personnel and technical co-operation among developing countries)

ESCAP energy projects such as REDP, PEDP, BSW (including the services of a solar energy expert) and RNSHP have been incorporated with TCDC elements and modalities at the time of formulation. Such TCDC projects provide participating countries with an opportunity for exchange of personnel, expertise and information in the context of TCDC. In January 1987, BSW implemented a TCDC-related research, development and demonstration training course in Indonesia on the evaluation, design and implementation of photovoltaic systems in developing countries, and a similar course is being planned in Pakistan.

(i)   TCDC working groups

A broader project will be formulated incorporating TCDC working groups as a mechanism for stable co-operative arrangements and covering the new and renewable sources of energy fields: solar photovoltaic, biomass combustion, including cooking stove technology, biogas, wind and geothermal energy. The TCDC working group mechanism is envisaged as self-sustaining intercountry co-operation.

On the basis of the nature of expertise involved, six TCDC working groups have been proposed, on: (i) electric power; (ii) natural gas; (iii) coal; (iv) energy conservation; (v) energy planning; and (vi) rural energy planning and development. These working groups will consist of professionals coming from the same functional areas. Because of their professional competence, they will be able to con-

tribute to the common objective of enhancing mutual benefit through regional co-operation in training, exchange of information and the use of existing regional facilities.

The secretariat will provide institutional and technical support to the TCDC working groups when they have been established. The groups would play a leading role in project planning, formulation and implementation. Their functions would be different from those of the focal points, which are mainly concerned with legislative work. A project proposal would be prepared by the Natural Resources division of ESCAP on the advice of such TCDC working groups. The groups would meet once a year to discuss common problems and develop co-operative programmes. In-country training programmes would be the main emphasis of the project. Resource persons from other developing countries should be encouraged to participate on a TCDC basis in the national training programmes.

Such regional TCDC arrangements could be extended to the interregional TCDC context. The promotion of interregional TCDC/ECDC by the regional commissions was accorded the highest priority by the Economic and Social Council at its second regular session in July 1985. The ESCAP secretariat has been identified as the lead Commission in the formulation of concrete project proposals on interregional TCDC in new and renewable sources of energy.

Regional commissions, together with countries of the region and certain expert bodies, have devoted substantial resources to developing institutions and implementing programmes to accelerate the application of new and renewable sources of energy. The depth of experience and institutional development in particular fields, however, varies significantly between regions. Interregional TCDC provides a rapid means of transfer of technology and experience between institutions in the regions, and thereby accelerates the application of such energy sources world wide. ESCAP is organizing the first operational interregional TCDC training-cum-workshop on biogas technology in China in mid-1987. An interregional TCDC/ ECDC project proposal on new and renewable sources of energy has been prepared by the ESCAP secretariat for submission to other regional commissions for possible co-operation on an operational TCDC basis. Interregional co-operation is already being achieved between ESCAP and the Economic and Social Commission for Western Asia (ESCWA) through ESCAP participation in the ESCWA intergovernmental meeting on new and renewable sources of energy scheduled for November 1987.

(ii)   Networking arrangements

The concept of networks is now accepted as a suitable approach towards better regional and international co-operation. Networks are basically associations of institutions for joint studies and exchange of information. In the ESCAP region, in the energy fields, at least six networks are either in operation or are envisaged to be formed. These are: (i) APENPLAN: (ii) Regional Co-operative Network for Energy Conservation; (iii) Co-operative Network for Rural Energy Planning; (iv) Regional Network for Small Hydro Power; (v) Regional Wood Energy Network; (vi) Biomass, Solar and Wind Energy Network. The concept involves a rotating lead institution co-ordinating the affairs of the network with all participating institutions or organizations co-operating with one another directly as well as through the lead institute that would facilitate and monitor arrangements. In some cases, some support from international organizations or donor countries may be expected at the initial phase, with the expectation that the network will ultimately evolve into a self-sustaining mechanism. At present networks (i), (iv), (v) and (vi) can be considered to be operational networks although sometimes with a limited range of activities, while the others are in the formative stage.

(e) Summary: Prospects for economic development based on human resources development in the energy sector (By the year 2010)

In view of the difficulties to be encountered in meeting future world energy demand, technological developments will be increasingly necessary to maintain sufficient energy supplies. Alternatives include both new technologies and existing technologies that become economic as energy prices rise. Recently, attention has been given to developing technologies for the better harnessing and more efficient use of energy resources, especially new and renewable sources of energy. As a result, human resources development has become an important component of the national economic policy of both developed and developing countries.

A great deal has been said about the potential and role of renewable sources of energy in meeting part of the future energy demand. In the developed countries, strategies for the exploitation of renewable sources of energy constitute a part of recent policies which aim at reducing the dependence on fossil fuels. Solar technologies (especially solar heating for buildings and for some industrial process heating), geothermal technologies, production of fuel from biomass, etc. are becoming more and more common in several developed countries. A particularly important social and economic advantage of the development of renewable sources of energy is the potential for promoting development in rural areas of the developing countries.

Each mode of energy conversion may penetrate the market at a different time, with a varying implementation and success rate. Among the technologies expected to be competitive in the future (in the year 2010) are heat pumps, enhanced recovery of oil, and the development of some new and renewable sources of energy. Of the renewable technologies, solar photovoltaic conversion, biogas technology, wind power, and mini hydropower are at present the most promising technologies for the future.

The application of these appropriate technologies will require substantial manpower, with suitable skills and qualifications. The principal justification of this new and active manpower programme in national economic development is that it would enhance the efficiency of macro-economic policy management: (a) by raising the productivity of the workforce through increasing the supply of educated and trained manpower; (b) by reducing structural and functional unemployment, the trade-off options between inflation and employment would be markedly improved. As a result of these two effects, manpower development policy could improve the efficiency of macro-economic management, and ultimately promoting a higher GNP (gross national product) growth rate.

5. Energy manpower planning, training and related socio-economic issues: an ILO perspective*

Introduction

In the past decade, manpower planning has not always been at the core of energy sector development efforts in Asian countries. Presently uncertain development prospects and energy policies are making manpower planning an even more difficult task. Unknowns surround not only the supply, demand and prices of energy resources, but also the very projects for which manpower had been trained. Major investments are delayed as a result of economic recession. Production processes change because alternatives are sought for energy-intensive products and by reason of energy friendly technologies. The rate of innovation in production technologies has increased. New energy resources are being developed. Manpower planners are thus faced with a greater variety of skills and more flexibility in utilisation. This requires an adequate response through training and other supplies. One problem is that the methodologies used by manpower planners are often not geared to such situations. Techniques based on fixed coefficients for skill inputs are inadequate. It has also proved difficult to relate the manpower requirements of the energy sector to national manpower programmes and labour market conditions. In these circumstances, it is useful to review issues and options in manpower planning as related to energy sector policies for the 1990s.

Human resources policies should be considered in the light of a longer or medium-term perspective on economic development. As other papers prepared by the ESCAP secretariat for its Committee on Natural Resources indicate,

---

* The paper is based on material prepared by the ILO's Employment Planning and Population Branch, and ARTEP, ILO's team for Employment Promotion in Asia who collaborated closely with ESCAP/REDP in manpower and training needs assessments for the energy sector; on ILO's programme in respect of social and economic effects of petroleum development and on other contributions (e.g. E. Lee: Economic restructuring and human resource development, APDC/ADIPA, 1986).

the uncertainty of the oil market continues to cast its shadow over major investment plans in industry and infrastructural sectors, including energy. The 1990s will witness further structural change, in the economic structure of Asian countries as well as in their energy sectors. There is, however, no doubt that economic growth will be accompanied by increases in overall energy consumption. Some Asian countries have elasticities, of energy consumption to GDP, which are still well in excess of one. It will require innovative energy demand policies to lower this ratio to, for example, 0.75, while attaining economic growth rates in the range of 5 to 8 per cent per annum. Within the energy sector itself, structural changes relate to the issues of substitution between sources and to the introduction of alternative supplies in rural areas.

Policy-making in human resource development has also felt the impact of the recession in the world economy. Programmes for restructuring have led to changes in skill-mix, to reconsideration of the content of training and to transferring part of the responsibility for training to the industries themselves. In many industries, in particular in manufacturing, this has meant a redeployment of skills, training for new growth areas and more generally a push for higher quality manpower. In infrastructures, the emphasis in the foreseeable future will be on better maintenance of the costly capital outlays. In service sectors, the existing drive for higher productivity will probably gain momentum. The absorption of excess manpower will fall on the agricultural and informal sectors. The next decade will demand higher levels of productivity and the energy sector will be in the forefront of this movement. Innovative labour market policies and training policies will have to accompany these developments.

The dimension of the development effort in the energy sector is large. The Asian Development Bank estimates in its study on this sector that countries will have to reserve 3 to 4 per cent of their gross domestic investment. Therefore, simultaneous development of manpower resources become important. There is thus every reason for the ILO and other agencies to draw attention to the neglected "human dimension" of the energy problem. A technical meeting, organised in 1982 under the auspices of ESCAP/REDP/ILO/UNESCO, reviewed experience in energy sector manpower assessment and planning. On the basis of the recommendations of this meeting, activities under REDP were oriented towards (a) assessment of manpower requirements of specific energy technologies by sub-sector, activity and occupation; (b) identification of relationships between manpower requirements by occupation and selected physical indicators (such as size, capacity, output, number of productive units, costs, etc.); and (c) formulation of manpower development strategies. This approach to energy manpower analysis was strongly influenced by ILO's experience in co-operating with the Government of the Philippines in this field. Detailed assessments were thereby made of manpower inputs required for different energy technologies. The focus was

clearly on sub-sector activities. Other exercises, for instance as executed under UNDP/World Bank Energy Sector Assessment Programme, tended also to emphasise detailed sub-sector analysis and reinforced the technology-engineering method to manpower planning.

A new dimension was introduced by ILO/ARTEP (Asian Regional Team for Employment Promotion) through its collaboration in the REDP. It attempted to go beyond the work already done on manpower requirements for specific energy technologies and sub-sectors by examining the issues and problems involved in formulating comprehensive manpower plans for the energy sector as a whole and in linking this to national manpower plans and policies. While the sub-sectoral and technology-specific work done so far or at present under implementation represents a considerable advance, it was obvious that there was a need to integrate these into overall sectoral as well as national manpower plans and programmes. On the demand side there was a clear need to aggregate the individual sub-sectoral forecasts of manpower requirements, to ensure that they are undertaken on a common basis. Only then can the full manpower implication of national energy plans be appreciated and the problems of manpower supply which are likely to emerge be identified. It is also essential to place the demand forecasts in the context of national manpower forecasts in order to identify competing demand for similar categories of manpower by other sectors. On the supply side similar arguments apply. Policies to ensure that the manpower requirements of the energy sector as a whole are met will need to be guided by a more comprehensive picture of demand than sub-sectoral or technology forecasts.

Labour market analysis is also important. Ensuring that the manpower requirements of national energy plans are met involves more than quantitative forecasting of demand and planning adequate supplies. The functioning of labour markets, and the wage and labour policies of employers in the energy sector will also be significant determinants of whether or not manpower requirements are actually met. These factors are particularly important for the energy sector because it includes projects in remote locations, those which involve large, lumpy investments with an irregular time-profile of labour and skill requirements during different phases, and those in which the disutility of labour is high. These aspects will have to be taken into account together with the more traditional aspects of manpower and training policies. In addition, other socio-economic effects might influence the shape of human resources development programmes. In the following paragraphs, illustrations are given of these aspects of planning in the energy sector.

(a) Manpower planning issues

The need for a new approach in manpower analysis has emerged from the country case studies, manpower profiles and related ILO/ARTEP work. This approach concentrates on human resource constraints in relation to specific development issues in the energy sector. The methodology aims at identifying critical shortages (or surpluses) of skills which exist or are likely to develop in the future, followed by recommendations to policy makers on how to deal with such shortages. The methodology is thus issue-oriented. In addition, while strengthening the analytical base of manpower and training needs assessment, it also promotes the establishment of longer-term capacity for continuous manpower analysis. It aims thereby at contributing to development options and policies in the energy sector. This new orientation focuses thus on three aspects: (i) the identification of manpower issues, (ii) the development of adequate methodologies and techniques, and (iii) the establishing of longer-term capacity of energy manpower analysis.

Manpower shortages, wastage and other personnel problems, as well as difficulties in preparing manpower for implementing costly expansion programmes, have been traditionally the main issues in energy sector manpower planning. These problems are limited to sub-sector priorities or reflect the urgency of energy development plans after the oil crises. It is only recently that other issues have emerged. Economic restructuring problems have accentuated concern about low productivity. This has led to promoting efficiency in the utilisation of scarce resources. There is now also considerable pressure to improve the returns on existing capital structures in conventional energy sub-sectors. Maintenance of facilities has to be strengthened. For new and renewable energy sources there is closer scrutiny of inputs and of results from extension work. Manpower issues today include a search for flexibility and a capacity to adjust to change and external shocks. Problems have to be solved more quickly, so as to minimise the risks to expensive capital equipment and installations. It is also increasingly recognised that energy programmes often have serious socio-economic implications, with close linkages to manpower, training and institutional issues. The list of manpower problems in the energy sector is thus rapidly expanding. However, it is felt that neither the present manpower planning techniques, nor the manpower development and training actions are adequate to deal with these issues. The first step in energy manpower analysis is therefore to identify the issues and technical manpower constraints for the development of the energy sector.

(i) Conventional energy resources

ARTEP's manpower planning and training needs assessments for the energy sectors of Indonesia and Pakistan are illustrative of the issues encountered in human resource development. Both countries invest heavily in their energy sectors. Pakistan, as a non-oil exporter, accorded the highest priority to the energy sector by allocating more than 38 per cent of its development outlay (1983-1988) to it. The constraints which energy places on the economy have been changing in nature. The supply side is rapidly becoming more diversified, with the domestic

oil-resources coming on stream and with planned addition of hydro electricity. However, energy conservation remains a major bottleneck. In the light of the energy constraints, the Sixth Plan contains measures to pursue growth in a more energy efficient manner. Inter-fuel adjustments are proposed to minimise impact. The Plan announces the development of indigenous resources of energy, including nuclear and renewables and the acquisition of technology relating to energy substitution. A programme of rural electrification to cover the entire rural population residing in compact villages has been announced. However, in practice, for the Sixth Plan, power was allocated 75 per cent of the investment resources for energy, mineral fuels 23.6 per cent and for renewable energy only 1.4 per cent. A further specification of the investment resources for the energy sector shows *de-facto* priority areas which manpower analysis cannot ignore.

It is interesting to note that skill distributions in energy sub-sectors differ considerably among countries. Total employment also varies over a wide range. Even countries with similar overall employment elasticities might show quite different (sub) sectoral rates. For example, both Indonesia and Pakistan have growth elasticities of employment of approximately 0.4. However, Pakistan has much lower rates than Indonesia if only the electricity sector is taken into account (0.23 vs 0.52). A more detailed analysis brought to light that Indonesia's electricity supply industry has a shortage of certain types of skill, graduates in particular. There is an absence of middle level personnel and a mismatch between categories of technical staff. The study on Pakistan, on the other hand, showed a certain degree of over-staffing in relation to its present level of electricity operations. In both cases, there is a growing awareness of the need to pursue higher efficiency in energy sub-sectors.

The country studies have assembled valuable and often new material on manpower in the energy sectors of Indonesia and Pakistan. However, the researchers were confronted with a lack of comprehensive reviews of energy sector manpower. There were insurmountable barriers for a full account of human resources in the energy sector, its planning, development, allocation, utilisation and monitoring systems, procedures and outcomes. Firstly, and most importantly, the teams faced the problem of the lack of a comprehensive energy policy. Despite the constitution of energy departments since the oil shocks of the 1970s much remains to be done towards the integration of energy programmes. Obviously there is a dual energy system, which is (inadequately) described as commercial versus non-commercial, or productive versus household fuels, or similar distinctions. What is clear, however, is that probably no Asian country can pretend to have an energy policy if it ignores the problems in cooking, heating and related energy uses of poor households. In addition, energy policies have become so entangled with wider socio-economic development issues that a clear framework is the excepton rather than the rule.

Secondly, the teams encountered difficulties in distinguishing between overall manpower aspects and specific manpower issues in the energy (sub) sectors. There are two sides to this problem. Obviously, energy manpower programmes cannot be treated in isolation from what happens to human resources elsewhere in the economy. However, there is also a lack of integration within the energy sector itself. There are strong centrifugal forces and sub-sectors tend to go back to the do-it-alone path. In so far as this obstructs the flow and retention of valuable skills for the energy sector this tendency should be avoided. In any case, it constitutes a bottleneck in energy manpower analysis.

Thirdly, and related to the first two problems, the teams were handicapped in not finding a central interlocutor. This reflects partly the lack of organisation in the sector. But it also points to the absence of an energy manpower unit which has intelligence functions and acts as a focal point for human resource issues in the sector. Inevitably, the lack of data and other information also posed severe problems for the study teams. To some extent this is a resource problem and with more time, the researchers would have been able to provide a better basis for their findings and conclusions. However, it reflects also the lack of interest in knowing what the manpower contribution is and for monitoring developments in the sector. Information leads to insight and more reliable policies. This in itself is sufficient reason to stress the longer-term concept of manpower assessment and monitoring. Finally, it should be recognised that the study teams came across a problem of analysis. Very few of the respondents have an analytical view of the manpower problems encountered in their (sub) sectors. The issues are not clearly identified and options for remedial action are rarely systematically reviewed. Manpower tends to be seen as a problem of inputs, more or less dealt with at sub-sectoral level. There is no strategy for examination of alternative production and personnel policies.

(ii) Household energy needs of the poor

Even in countries with highly developed commercial fuel systems, the majority of the population has to rely on biomass from the nearby environment. For example, in Pakistan more than 70 per cent of the population depend on firewood, dung cake, farm residues and other biomass resources. It would seem that the official estimated share of approximately 30 per cent in total energy consumpion is an underestimate of the energy crisis which poor households in particular are facing. The manpower requirements for meeting this challenge are not well known. Staffing problems appear as soon as large commercially based operations have to be transformed into smaller decentralised and local-community oriented services. The introduction of new technologies is hampered by the lack of social engineering skills requied for large-scale dissemination.

It is increasingly recognised that improving the energy resources of the poor requires not only changes in investment priorities within the energy sector, but also different human resources policies. This can be illustrated by some of the characteristics of household energy development programmes, in particular when new and renewable sources of energy (NRSE) are compared with electric power supply. In the case of electricity corporations, capital intensive structures are maintained by a highly skilled workforce. The technology and skills required are well known. Wages do not constitute the major cost category in production. Training programmes can be established in relative isolation from other sectors in specialised training centres with adequate financial and staff resources. It is quite possible, when temporary shortages of skills occur, to use expatriate skills, as these can easily be fitted into the universal technology and manpower utilisation patterns. Quite the contrary is true for new and renewable sources of energy. Production, transformation and utilisation of these fuels take place in a variety of circumstances, which often leave little room for direct intervention by policy-makers and management. Production and consumption can occur in the same household without any capital or market intervention. Technical solutions might be well known, e.g. for forestry, but skills are lacking to promote implementation by the masses of the population. In other cases, technologies, e.g. solar, are introduced which are neither technically mature nor acceptable in the present socio-economic conditions of developing countries. In particular for household fuels, there is a need for skills in social engineering. The key to the further development of this sub-sector is the integration of demand in more effective energy systems at macro as well as village and household levels. This requires enhanced attention to the social dimension of energy production and utilisation. In terms of energy supply, renewable sources of energy frequently require decentralised production operations with relatively heavy inputs of supervisory and management skills. Personnel resources will take a large share of the cost of development programmes of NRSE. Training programmes are often close to those of existing trades (e.g. plumbing for solar installations) and thus there is competition for skilled manpower. Successful introduction of NRSE is dependent on application of technology and effective dissemination, requiring skills which are in scarce supply. If the policy is to promote household energy availability, then manpower and personnel action will need to be adapted to this priority.

An example of an area where manpower analysis could play an active role is the supply of biomass energy. Most of the household fuel programmes have been approached in the typical "extension service" manner. The methodology was basically not different from the staffing of, for example, (industrial) forestry operations. Organisation and methods criteria were applied to well defined projects. Staffing norms were elaborated, mostly on the basis of density ratios. This "top-down" approach has not been very satisfactory due to lack of participation from farmers and energy users, and the financial and social costs of these programmes have been high. It is doubtful that systems which have proved to be useful for cash crops, such as the World Bank's T(raining) and V(isit) S(ystem), can be applied to household fuels in the traditional sense. Alternative development approaches for meeting (rural) household energy needs have different, though mostly unspecified, staffing implications. The various forms of participatory action have hardly come out of the pilot phases. One feature seems to be common to all the "alternative" or "bottom-top" approaches: training of the "change agents" is quite short. However, the approaches differ in respect of the general level of education required of these agents, which may vary from none (the contact farmer approach) to high (in the case of professional facilitators utilised, for example, in India's biogas programme). Obviously, good development agents have come forward through adequate (self) selection processes and which have on-the-job experience. Perhaps a successful formula could consist of the best elements of the T and V system (such as only one agent interfacing with the farmer and with adequate back-up of sector specialists) combined with the principles of the participatory approach. However, many gaps remain in knowledge of the household energy sector. Manpower analysis could be oriented to assist in remedying these weaknesses. It is also clear that in this sub-sector applied manpower research is still required and that contrary to the more conventional sub-sectors, indicators of present skill utilisation and labour market signals are less reliable.

(iii) Integrating manpower analysis with energy development planning

The organisation of the energy sector should reflect the main development issues encountered. Large countries, with important reserves in fossil fuels and developed commercial fuel systems will have different requirements than small countries with, for example, final energy consumption mainly confined to biomass in poor households. However, experience shows that in most cases poorly co-ordinated institutional structures prevail. In particular, energy planning and policy formulation tend to be weak. As a consequence, management of energy activities is left to sub-sector corporations. This leads to the lack of overall development in the energy sector and neglect of important socio-economic considerations. In this respect, it is useful to keep in mind the features of a basic three-tiered organisational structure, whereby a distinction is made between policy-making, planning and management. At the highest authority, policy formulation and strategic planning takes place. This is also the level where energy policy is integrated with national socio-economic development objectives. At the base are the various energy agencies, which carry out operations and execute the sub-sector programmes. These agencies should have full responsibility for execution of their projects and for management of inputs. They should be independent, within broad policy guide-

lines and planning instructions. Between these two levels of authority remains an important co-ordination function to be fulfilled. This is the role of the energy-secretariat. If there is a department which has the responsibility for planning energy production, conservation and price-setting, then a manpower planning base should be there as well.

Efficient resource mobilisation requires that manpower analysis be integrated with energy sector planning and development. It postulates a comprehensive approach to issues and options in the energy sector, including manpower, training and other social implications. The major prerequisite is that a policy basis or perspective exists for integrated energy sector development. Without an energy policy, based on assessment of end-use and supply sources, there can be no sectoral manpower analysis and planning, but only sub-sector exercises in this respect. Experience shows that shortages in the well-established energy enterprises (electricity, for example) or in those sub-sectors which can afford to buy the skills (private oil companies) are noticed and dealt with earlier than in others (forestry operations). In order to establish a sustained capacity for energy manpower analysis, an action plan will have to be followed. In the process, staff will have to be trained, a conceptual framework introduced, a planning routine established and diagnostic services developed.

Energy policies can have a profound impact on employment, income distribution and poverty alleviation. How this will work out depends on many factors, including regional considerations. The point is that the alternatives have different manpower implications which have to be assessed. For example, if the primary goal is to provide energy for increased agricultural production and productivity, in areas where the people/land ratios are high and capital resources scarce, then this requires an energy manpower policy responding to this priority. It will be necessary to promote forestry and agro-forestry, and the requisite skill formation including the acquisition of skills for conservation. In addition, such a policy would probably imply that the urban sector will have to switch to non-traditional fuels with wide-ranging effects upon skills. Thus energy policies have manpower implications. A major function for energy analysis is to promote co-ordination between national human resource policies and energy sector staffing programmes.

(b) Training and the utilisation of skills[1]

While manpower analysis identifies shortages or surpluses in skills in relation to energy sector development programmes, training remains a very important instrument for matching jobs and skills. If shortages exist, training can help overcome them through upgrading lesser skilled persons. If there are surpluses, these can be resolved

---

[1] Much of this section has been taken from Chapter 5 of *Energy Manpower Analysis, A Manual for Planners*, ILO/ARTEP, UNDP/ESCAP (New Delhi, 1987).

through retraining of redundant workers for other jobs. However, training is expensive and often not the most effective means for solving skill gaps. From a manpower analysis point of view, training might be one of the last measures to take, once all other means to narrow skill gaps have been applied. Therefore, it is the task of the manpower analyst to determine as precisely as possible the training necessary to provide for the right skills for the right job at the right time and thus to maximise the availability and competence on the job, in order to contribute to higher productivity and to minimise costs. Often the primary focus will be on training provided by the energy sub-sector industries themselves, to strengthen the utilisation of skills and to capitalise on the competence available in the production process.

This emphasis on training in relation to production processes is thus specific to establishment or work-organisation. Obviously, most forms of education and instruction have more general objectives than just filling skill gaps in firms. Training adds to human capital, thereby improving the employment situation of individuals. It also contributes to efficiency in society. In addition, important social objectives are pursued with training, such as lessening inequality. The point is, that from the establishment point of view, these wider objectives are considered external effects. They may be desirable and positive, but they should not be, as a rule, the main reasons for undertaking energy sector training.

The relevance of training, however, should be considered from the standpoint of the individual, the employing establishment as well as in a wider social perspective. The complexity of these relationships is a major problem for sectoral manpower planners, in particular for energy manpower analysts who are faced with a wide variety of training programmes. It will be necessary to narrow the notion of human resources development to a concept which is operationally relevant to manpower utilisation, labour market functioning and economic forecasting. A common denominator, but one which can be quantified, has to be found. In what follows, it will be argued that "jobs" may be chosen as the subject of analysis.

Adding new qualifications, relations and values to the human element in an organisation has meant a change from traditional personnel administration to the wider concept of human resource management. However, in relation to performance at the workplace, the level of execution is still measured by concepts of skill. The dimensions of skill are information, knowledge and competence or know-how. All three dimensions change continuously in value, not only for individual skill holders but also among tasks to be executed. The relationships between skills and tasks are influenced by combinations of production factors and variations in output. Training for skills means an organised transfer of continuously changing information, knowledge and competence. The impact of training on production varies also from increased output, to decreases in inputs

or changes in the quality of performance. The complementarity of higher skills and capital, as well as the substitution effects between different types of labour, are additional indications of the many conditions which surround the real effect of training, as measured at the workplace. It may be concluded that training cannot be considered in isolation from other factors determining performance of the job.

A training needs assessment is a review of the nature of training in the context of a specific learning situation. The latter is closely related to job-skill gaps and supply analysis. It will therefore always concern a trainee, a production process and a training programme. The bridge between these three aspects is the job, i.e. a specific combination of tasks in production. The experience of ILO's (vocational) trainers is that manpower planning, labour market information, occupational analysis and classification provide the foundation not only for training needs assessment and vocational training planning, but also for curriculum development and for monitoring internal and external efficiency of training programmes.

This practical experience of vocational training planners has helped to bridge the gap between macro (employment) planning, sectoral and micro considerations, labour market analysis and training programmes. The following building blocks may be added to the concepts of manpower analysis, training analysis, training assessment and training needs assessment:

(i) Field of work: a unit of analysis which can be applied at sectoral and micro levels;

(ii) Occupation: the type of work performed constitutes a classification which is useful to both macro and labour market analyses;

(iii) Job: a set of specific tasks, which is the basic analytical concept for employment in the production process; positions are descriptions of differences in the levels of responsibilities and other characteristics of the work;

(iv) Task: a unit of work to be accomplished; task elements constitute the specific activities required;

(v) Learning element: a package for acquisition of information, knowledge, and competence to execute a task, transmitted in a manner such as to motivate the future workers;

(vi) Trainee assessment: specification of the capabilities of the trainee in relation to the learning element, as well as of his attitude to work.

A training assessment builds a bridge between the trainee, the work (including other factors of production) and the training programme. Job analysis is the pivot in this whole concept. It provides the basic materials not only for employment, labour market and training planners, but also for vocational guidance and placement. On the basis of a detailed examination of a job, standards can be established for training and performance.

Efficiency of training programme

Training can take many forms. Usually three categories are distinguished:

(i) off the job, in schools, special centres or in establishment;

(ii) on the job, both formal and informal;

(iii) mixed, a combination of (i) and (ii) including also learning-cum-production schemes, such as apprenticeship.

Work experience is a special form of the training situation, which may be organised such as in rotation schemes for high level personnel. The evaluation of these schemes is a specialised assignment which goes beyond the normal duties of an energy manpower analyst. However, recent literature in training assessment brings out the relevance of close co-operation between manpower analysts and training planners. Their common ground is a job analysis, a concept bridging occupations and training programmes.

Training programmes which respond to such job analysis require not only suitable learning materials and other resources but also a constant adaptation of the training environment. In order to increase the flexibility of training, to reduce costs and to strengthen the linkeage with the work place, new training methodologies have been developed. The ILO's systems approach to vocational training led to the development of "Modules of Employable Skills" (MES) which is applicable to the energy sector. In the case of supervisors, the adequacy of training is more difficult to determine. Instruction for supervisors is a special type of management training, and the ILO has also developed special modular training materials for these situations which are also directly applicable to the energy sector.

Manpower analysts and training planners are both interested in the efficiency of the training provided. Two aspects can here be distinguished: internal and external efficiency. Internal efficiency is limited to the training process itself and concerns the rationality of processing the trainees. External efficiency refers to the outcome of social costs and benefits of the programme taking into account employment of the trainees. One should always ask the question, are there other ways to achieve manpower objectives which would be faster, easier, cheaper and less cumbersome? It is also necessary to consider both quantitative and qualitative aspects. However, obviously external efficiency is more directly relevant to the energy manpower analyst. The evaluation of the external efficiency of a training programme is often done along the lines of a cost-benefit analysis[2] which, while subject to a number of caveats, still provides some interesting results, for example, the following points may be relevant for an energy sector training programme:

(i) shorter courses are often more efficient than longer ones, in particular in vocational training;

(ii) a mixed form of training continuing off-the-job and on-the-job training is more efficient;

(iii) internally efficient training programmes tend to score better also on external efficiency;

(iv) work and labour market experience immediately after training is not necessarily a good indicator of long-term benefits;

(v) external effects of training can be substantial: these include better use of other resources, but also potentially negative aspects such as migration and displacement.

For measuring the external efficiency of training programmes, it is essential to do follow-up studies on trainees. A tracer study measures the effect of a training programme. It consists basically of questioning students, trainees, graduate workers and employers/supervisors about the linkage between training and the world of work. Tracer studies can have a number of specific objectives and are normally very resource demanding in organisation, technical experts and finance. They are often used to check on two specific aspects:

(i) the degree to which training is actually utilised in the job; and

(ii) the supervisor's assessment of the skills imparted by training and the impact on productivity.

It is very important to improve channels of communication and in this regard tracer studies may be a particularly useful tool.

While this section on training has been particularly focused upon vocational training, it should be stressed that supervisory and management training requirements are of equal importance to the success of energy programmes and projects. This is particularly true as regards rural energy activities where it will be often necessary to combine vocational, supervisory and management skills within a single employee in light of the development of decentralised and small-scale energy installations.

Attention should also be drawn to the importance of the training of trainers and the development of appropriate incentives to keep them within training institutions. The development of qualified trainers and their conditions of

---

2 For a detailed review of cost-benefit analysis and rate of return analysis see *Energy Manpower Analysis* (pp. 87-91).

work and wages are important issues which deserve special attention in the energy sector.

(d) Manpower and other socio-economic issues in petroleum development

The energy sector is of course not uniform, and it may be helpful to consider a particular sub-sector — in this case, that of oil and gas — in order to see concretely what form energy manpower planning and training issues can take. It is not difficult to justify paying special attention to this sub-sector, given its economic importance.[3] The first point to make is that policies at the enterprise as well as national level need to be considered when assessing manpower planning and training efforts, since the petroleum sector is characterised by big firms, which both can and must make their own provisions for obtaining skilled manpower. That tends to be true irrespective of the relative share of state-owned and private enterprises in a particular country.

For example, in India, where virtually all oil/gas activities are in the public sector, the approximately 90,000 persons employed in the industry in 1984 were distributed among only twelve organisations (43,435 employees in two upstream firms, and 47,375 in ten downstream companies)[4]. In Malaysia, where the industry is mixed, state-owned Petronas had 5,742 employees, whereas of the largest privately-owned groups Esso had 2,488 and Shell (upstream only) 2,871.[5]

Enterprises of this size, technological sophistication and capital-intensity require comprehensive systems of recruitment, training and succession planning to ensure their supply of skilled personnel. How much they provide themselves depends on what the particular national educational system offers, the availability of relevant skills on the local labour market, and government requirements — incorporated in national law and/or in the contracts governing their petroleum operations — with respect to indigenisation, minimum training expenditures, preference for local labour, etc. At one end of the spectrum, where the university system does not have programmes in some of the relevant disciplines — such as geology, geophysics, petroleum engineering — companies may go so far as to support or even initiate action to establish and nurture such departments, e.g. by endowing a chair. At the other extreme, in countries where the universities offer higher education in the relevant professional disciplines and

where vocational training institutions supply adequate numbers of skilled labourers, artisans and technicians to the local labour market, the companies can take a much narrow view of their role, concentrating on on-the-job training and on internal personel systems needed to retain staff, to offer normal career progression and to keep pace with evolving technology. While the details of this intervention vary considerably from one company and country to another, the basic point is that the enterprises not only influence the demand for skilled and professional manpower; they also may play an active role in stimulating the supply.

Investments in human capital — not only through training but through compensation systems aimed at retaining and motivating staff over long periods — tend to have a long payback time. For commercial enterprises — including state-owned enterprises run on commercial lines — to make such investments on a large scale requires financial strength and an ability to take the long view. These are not characteristics of all sectors, but fortunately they are, by and large, features of the petroleum industry.

At the level of national policy, governments must determine as a function of their petroleum resources and consumption how extensive their need for skilled manpower is in this sector. All countries, at a minimum, require some expertise in international products supply, in internal distribution and in evaluating the country's geological potential. These skills are of course available on the international marketplace, but effective negotiation with would-be providers requires a minimum knowledge within the country of how the industry works. Beyond that minimum, needs and resources vary considerably from country to country. Most countries have some refining capability, and hence find it appropriate to offer in their systems of higher education professional training in the relevant disciplines, such as chemical and mechanical engineering, all the more so as those specialities have potential application going far beyond the petroleum industry. Fewer but still many countries have proven petroleum reserves, or the hope of finding some, and therefore support programmes in universities and technical colleges in petroleum-specific disciplines such as petroleum geology, petroleum-related geophysics and petroleum engineering. Those that have substantial petroleum production will probably have a broad range of such programmes, capped perhaps by a national petroleum institute to serve as a kind of custodian of the national skill base in the sector. Since these programmes are relatively expensive, this approach is normally justifiable only in countries with relatively advanced, integrated petroleum sectors, or the determination and means to acquire the same.[6] Alternatively, *regional* institutions and programmes may make

---

[3] An indication of that importance is the fact that oil and natural gas represent a high proportion of Primary Commercial Energy in the region (e.g. Japan, 65 per cent; Australasia 49 per cent; South East Asia 70 per cent).

[4] Tata Energy Research Institute with C.R. Jagannathan and Charit Tingsabadh, *India and Thailand: Social and economic effects of petroleum development*, (Geneva, ILO, 1987).

[5] Peter Hills and Paddy Bowie, *China and Malaysia: Social and economic effects of petroleum development*, (Geneva, ILO, 1987)

[6] These and related issues are considered in Jon McLin, *Social and economic effects of petroleum development in non-OPEC developing countries: Synthesis Report* (Geneva, ILO, 1986)

sense where conditions do not justify such investments at the national level.

Also appropriate for consideration at national level is the adequacy of the numbers and output of workers with intermediate skills — artisans, technicians, etc. In most developing countries, bottlenecks are found at this level rather than that of professional expertise. Since most of these skills — welders, electricians, computer technicians, etc. — are not specific to the petroleum industry and can thus be transferred between sectors, the case for a national effort to promote them — as an adjunct to petroleum training or independently of it — is very strong.

The petroleum industry is international by tradition. One thing this has meant is that, as an alternative to comprehensive indigenous development, small and less developed countries could have recourse to foreign investors to exploit their petroleum resources or meet their needs for refined products. Whereas this option tended formerly to be an all-or-nothing affair, the diffusion of petroleum expertise in the last quarter-century or so has created a situation in which governments can now more or less choose what part of the sector they wish to manage entirely through national resources, and can contract for the remaining areas of equipment and/or expertise in the international marketplace.

While that is a comforting option to have, countries with a growing resource base typically wish to enlarge the share of their needs — both in terms of expertise and equipment — which they meet from their own resources. Requirements concerning local preference, indigenisation and training obligations are now commonplace in production sharing contracts negotiated between foreign investors and host governments. As a result of the efforts made in these areas by governments and private operators, the role of expatriates in the industry is much reduced compared to the situation of fifteen or twenty years ago.

One implication is that technology is now diffused according to a somewhat different pattern. Previously, expertise was largely disseminated between affiliated companies belonging to the multinational groups that typified the industry. While that still occurs, the industry is much more fragmented than before. Even local operating companies belonging to multinational groups have more of a national character than used to be the case. (Already in 1985, Exxon pointed out that 98 per cent of its employees were nationals of the countries in which they worked. Since then, the group has made major reductions in its workforce, and these fell most heavily on expatriates and U.S. staff; hence the proportion today is probably even higher.) And the larger role played by state petroleum enterprises and by smaller companies means that a lot of training that used to occur through the international movement of personnel between affiliates now takes place as part of a contractual arrangement between firms operating entirely on an "arms length" basis. One question

which might deserve attention is whether these arrangements are effective in allowing national enterprises to keep their staff fully abreast of technological and commercial developments occurring in other parts of the world petroleum system, or whether they need to be supplemented by new mechanisms facilitating the international movement of personnel among national oil/gas companies. Is there a potential role to be played here by the regional association of petroleum enterprises such as ASCOPE[7] and its counterparts in other regions?

Many of the points made above can be exemplified by considering as examples two ESCAP countries, India and Malaysia, on which the ILO has recently done work. In India, where the industry is virtually completely in the public sector, major indigenisation and training efforts have gone hand-in-hand over the last quarter-century or so. The rising share of petroleum needs met from indigenous sources (from 6 per cent in 1970 to 70 per cent in 1984) can be taken as an indication that this effort has produced positive results. In 1984, only 140 of the 43,435 persons working in India's upstream sector, and none of the 47,375 working in the downstream, were expatriates. This process has been supported by a massive training effort of which one recent example is the decision by the Oil and Natural Gas Commission (ONGC) to establish a Training and Executive Development Institute in Dehradun, which is to hold short trainning courses for about 3,000 executives per year. At the vocational level, three Staff Training Institutes are to be set up during the current five-year plan period which will train about 6,000 workers and supervisors per year.

In Malaysia, the ownership pattern is different — a substantial share of the industry remains private, and the state company Petronas is both a joint venture partner of some of the private firms and an operator in its own right. But the indigenisation and training story is similar. Between 1973 and 1986 the ratio of expatriates to Malaysians among senior staff in the industry was roughly reversed: from 3:1 to 1:3, and that in a group which in absolute numbers was expanding rapidly. The largest private upstream operators, Esso and Shell, and Petronas all spent very heavily on both in-country and overseas training in support of this effort, which involved very large numbers of staff. And oil production increased over that same period from 90,000 b/d to about 500,000 b/d.

As the above brief review of some issues in the petroleum sector points out, in addition to general energy sector manpower planning and training issues, it is essential to also look at special sub-sector requirements in order to design and implement an effective and efficient human resources development programme in the energy sector.

ASCOPE — ASEAN Council on Petroleum

(e)   The ILO approach

In view of the growing employment, training and social implications of the changing world energy situation, the ILO is increasingly active in the energy sector. The ILO has identified three related and complementary approaches for the future development of its programme of energy-realted activities:

(a) ILO action to assist governments, employers and workers to develop their capacity to deal effectively with the labour and social aspects of the energy situation (DIRECT ACTION);

(b) ILO action to advise and support international and regional organisations (including ESCAP) in their energy activities as regards employment training and social aspects of energy (CATALYTIC ACTION); and,

(c) In the context of its overall contribution to the world of work, the ILO promotes the integration of energy activities with other aspects of socio-economic development (INDIRECT ACTIONS).

Careful attention is given to ensure that all ILO energy-related activities be within the ILO mandate and must conform, in their objectives, to the overall objectives of the ILO. In this regard, special emphasis is placed on energy-related projects which aim at: the alleviation of poverty; the satisfaction of basic needs; the creation of employment or other income-earning opportunities; the more equitable distribution of income; the development of vocational and management skills; and improvements in the conditions of work and life, especially for the rural poor and special categories of workers such as women and refugees. The ILO regards the improvement of human resources, in all its aspects (e.g. skills, knowledge, as well as physical and mental well-being) as the ultimate aim of socio-economic development. The planning, development and optimal utilisation of human resources are at the core of the ILO's work, including in the energy sector.

The present ILO programme of energy-related activities is giving particular emphasis to (a) manpower and training assessments in the energy sector (already described in some detail above), (b) training in the energy sector, and (c) social and economic implications of energy policies and programmes.

The ILO has a long tradition of vocational, management and rural training activities related to energy. Particular reference whould be made to ILO vocational training activities in the electric power sector in various countries, e.g. Malaysia, Nepal, Papua New Guinea, Philippines. As regards forestry and charcoal training, the ILO has carried out a regional workshop and seminar in Thailand in 1983 and several follow-up workshops in Burma in 1984 concerning fuelwood and charcoal preparation. In addition, the Office has revised its Training Manual on Fuelwood and Charcoal Preparation and is assisting governments to give training in this field. The ILO's programme of rural artisan training has also included programmes and activities related to training-cum-production for devices related to new and renewable sources of energy (NRSE). In addition, the ILO's Turin Centre has developed a wide-ranging programme of training courses in the energy sector including courses on energy management, management of electrical supply utilities, and NRSE. The Turin Centre's most important contribution has been the development of detailed training modules for energy management at the enterprise level. These modules have been developed and tested and the Turin Centre is now carrying out a series of training-of-trainers courses to facilitate the use of these modules in developing countries.

Following a rather rapid increase in the number of requests for ILO assistance for training in the energy sector, the ILO has recently established a special position of Inter-Regional Adviser on Energy Training. It is expected that a significant portion of his time will be devoted to the development of activities in the ESCAP region. His activities are being very closely integrated to the ILO's major efforts concerning energy manpower planning and training assessments. The ILO is providing assistance to governments, often in close collaboration with the World Bank and other international and national donor institutions, with a view to developing energy manpower planning and training projects linked to national energy plans, projects, and investments. The ILO would welcome any inquiries from governments concerning possible ILO assistance in this field.

6. Report of the Meeting of Senior Experts Preparatory to the Fourteenth Session of the Committee on Natural Resources
(E/ESCAP/NR.14/4)

(a)   Organization of the meeting

The Meeting of Senior Experts Preparatory to the Fourteenth Session of the Committee on Natural Resources was held at Bangkok on 14 and 15 May 1987.

Attendance

Senior experts from Bangladesh, China, India, Indonesia, Thailand and the Asian Institute of Technology (AIT) participated in the Meeting. A list of the experts is annexed to the report.

Opening of the Meeting

The Deputy Executive Secretary of ESCAP, in his opening statement, said that he attached great importance to the Meeting because it would consider in-depth important energy issues to facilitate the deliberations of the Committee on Natural Resources on those issues and also because the Meeting could provide valuable guidance to the secretariat in formulating its medium-term plan, 1990-1995 according to the needs of the countries of the region.

Noting the key role of coal, natural gas and electricity in the field of commercial energy supplies and their increasingly greater impact because of the resource limitations of oil, the Deputy Executive Secretary stressed the importance of addressing a number of issues to achieve the goal of low dependence on oil. Such issues included the building up of adequate infrastructure and institutional capabilities to facilitate an increased volume of diverse energy trade. To maintain a sustained supply of new and renewable sources of energy, which played a crucial role in meeting energy needs in rural areas, strategies were required to be adopted at both the national and regional levels. Human resources development for the expanding and increasingly complex energy sector, both commercial and traditional, was another critical issue in all the phases of planning, development, operation and maintenance of energy systems.

Finally, the Deputy Executive Secretary expressed confidence that active participation in a congenial, informal, "brain storming" atmosphere would bring fruitful results.

Election of officers

Mr. T.L. Sankar (India) was elected Chairman and Mr. Lu Yingzhong (China) Rapporteur for the Meeting.

Adoption of the agenda

The Meeting adopted the following agenda:

1. Opening of the Meeting.

2. Election of officers.

3. Adoption of the agenda.

4. Consideration of issues on prospects for production and utilization of coal, natural gas, electricity and new and renewable sources of energy.

5. Consideration of issues on human resources development.

6. Consideration of the medium-term plan, 1990-1995, in energy.

7. Adoption of the report.

(b)   Proceedings

(i)   Consideration of issues on prospects for production and utilization of coal, natural gas, electricity and new and renewable sources of energy

(Item 4 of the agenda)

The Meeting had before it document NR/MSEP/1 entitled "Prospects for production and utilization of coal,

natural gas and electricity." The Meeting also had before it two draft papers on "Updated assessment of the contribution of new and renewable sources of energy to regional energy supply" and "The co-operative research, development and demonstration achievement and future plans on new and renewable sources of energy". Three papers, presented by the experts from Bangladesh, China and Indonesia, were also considered by the Meeting.

Secretariat document NR/MSEP/1 dealt with the importance of large-scale energy systems in general and with the role of coal, natural gas and electricity in particular. It was suggested in the document that to meet the growing demand for high-quality energy, intensive developments of coal, natural gas and electricity systems (from production to utilization) should be undertaken, keeping in view the resource limitations of oil. The importance of international co-operation was emphasized in establishing integrated large-scale energy systems.

In the papers presented by the three experts from Bangladesh, China and Indonesia, some relevant issues were brought in regarding the production and utilization of coal, natural gas, electricity and new and renewable sources of energy. Those issues concerned the building up of national capabilities; technology transfer; integrated management of all sectors of the energy system; inter-fuel substitutions; pricing policies; the oil market situation; energy demand projection; fuel options for electricity generation; safety and environmental issues; energy trade; conservation and efficient utilization; loss reduction, and so on.

The draft paper on the assessment of the contribution of new and renewable sources of energy to regional energy supply covered the aggregate contribution and analysed regional programmes and trends of new and renewable sources of energy in contributing to the energy supply. It was shown in the paper that traditional energy supplies still contributed up to 50 per cent of the energy supplies of the developing countries of the region and were crucial and often the only available supplies for people in rural areas, who represented 80 per cent of the population in the region. To maintain a sustained traditional energy supply, several measures would have to be taken by the countries as well as by regional and international organizations. The paper reported that experience with efforts over the past one and a half decades had shown that biomass resources were unlikely to be able to replace conventional commercial supplies (e.g. oil, gas, coal and electricity) completely in providing for expanded rural needs. Thus, to satisfy regional energy requirements, a mix of renewable and non-renewable resources would be required. The challenge in rural energy planning was to keep a balanced perspective of an assured supply of energy at affordable prices.

The secretariat draft paper on co-operative research, development and demonstration achievements and future plans concerning new and renewable sources of energy reviewed the achievements of regional co-operation in research, development and demonstration on new and renewable sources of energy based on the activities conducted by the secretariat over recent years. Identifying the priority areas (solar photovoltaic, small hydro, biomass, fuel cell) for which further co-operative research, development and demonstration were of importance, future plans for co-operation were discussed. In order to promote regional co-operative research especially on new and renewable sources of energy, the concept of a regional centre for research and development was put forward, involving tripartite co-operation among regional research institutes, one of which would play the role of research centre, and developed and developing countries. For the first step in promoting the concept, it was recommended that a tripartite consultative meeting be convened among research institutes, potential donor countries and interested developing countries.

The Meeting found all the documents submitted by the secretariat and the experts useful in bringing out salient issues in the development and utilization of various energy sources. Some updating and improvements in the secretariat paper were suggested.

The Meeting deliberated intensively on various issues and made the following conclusions and recommendations.

Conclusions and recommendations

The discussions at the Meeting emphasized the need to consider the prospects of production and utilization of coal, natural gas and electricity on a systematic basis. The level to which the different fuel forms would be utilized would depend on the relative prices of oil, coal and electricity; those in turn would depend on the production costs in the countries concerned based on natural endowment as well as on the price of oil set outside the region. In effect, therefore, the price of oil on the international market would have a significant impact on the prospects for production of the different fuel forms.

It was also emphasized that most of the countries derived the forecasts of demand for fuels from the development plans which set out the desired rates of growth of the economy. The view was expressed that even such rates of growth would be dependent on the price of oil to some extent.

However, the Meeting recognized that while forecasting oil prices was not an easy exercise some assumptions had to be made purely for the purpose of deriving a broad picture of the total energy requirements as well as the relative share of fuels. After the discussions, it was assumed that the oil price currently prevailing could be taken as the

base and a steady but slow increase in the real price of oil might be registered.[1]

The Meeting also considered that changes in the international oil price would not automatically result in an adjustment of domestic oil product prices. There were advantages and disadvantages to oil price adjustments. Countries would follow the policy consistent with their overall economic (fiscal and monetary) policies of management. A review by the Meeting of the price adjustments made to the falling prices of oil indicated that most of the countries of the region would pursue strong energy conservation policies with special reference to oil. The price changes were unlikely to affect consumer prices but would have a significant impact on national development policies.

Energy demand

In the absence of detailed plans on a long-term basis by the planning agencies of the countries, the demand forecast for energy would continue to be based on informed guesses made by experts. The Meeting noted the estimate discussed at the Asian Forum on Energy Policy in October 1986, which broadly indicated an overall energy consumption increase of about 4.5 per cent per year in the countries of the region. This tallied with a similar rate of energy consumption growth projected for China.

It was pointed out that currently the elasticity of energy consumption to gross domestic product (GDP) in the region was close to 1 in many countries, and considerably more than 1 in some countries. As the desired rate of growth of GDP in most of the countries of the region was more than 5 per cent and closer to 7 to 8 per cent per annum in many countries, the energy growth rate of 4.5 per cent per year implied that very strong conservation efforts needed to be made by the countries of the region. The example of the country exercise for China confirmed that. The development plans of China assumed about 7 per cent growth per year of GDP but it was anticipated that the energy intensity would decline sharply leading to an overall energy/GDP elasticity of 0.6 or 0.7 by the year 2000.

The Meeting drew attention to the fact that even a 4.5 per cent overall energy growth rate would place a serious burden on the investment capabilities of the countries of the region. If energy conservation efforts, with special emphasis on oil conservation, were not rigorously

---

[1] Starting with the 1987 price of around $18 per barrel, it might move to $20 by 1990 and might increase at a rate of about half to one US dollar per year. The resulting price might be:

	Per barrel	
	2000	2010
Low-price scenarios	$25	$30
High-price scenarios	$30	$40
(All in 1985 prices)		

pursued, energy shortage would become a constraint to economic growth in most of the countries.

Coal production and utilization

The Meeting was of the view that by the year 2010 the region would become the largest producer of coal in the world. In view of their relatively lower oil endowment, China and India were likely to embark on plans for sharp increases in coal production and utilization. Most of the increased coal production would be used for power generation. Moreover, the coal-producing countries would use coal in larger quantities for industrial production, especially in steel and cement production, and also for household consumption, which might be increasingly in the form of gas. The coal-importing countries would mostly use coal for power generation and in limited quantities for steel and cement production. Australia, China, India, and Indonesia would increase their coal production. There was likely to be a big coal market in the region, with the increase in demand for imported coal from countries like Japan and the Republic of Korea. The ESCAP studies on coal made under the regional energy development programme (REDP) in the years 1984-1987 were broadly endorsed by the Meeting. Even at the modest rate of growth of coal production increasing at around 4 to 5 per cent in China, coal production would reach 1,300 million tons by the year 2000 and 2,200 million tons by the year 2010 for China alone. In India, coal production would be increased to over 400 million tons by the year 2000 and over 600 million tons by the year 2010 from its current level of 165 million tons. In Indonesia, it would rise to 50 million tons in the year 2000 starting from the current level of less than 2 million tons.

The Meeting felt that in view of the past performance of the coal industry in China and India, the projected growth rate could be achieved if adequate funds were available. But the major constraint would be the transport of coal (a solid fuel) for both China and India, which were large countries where coal reserves were concentrated in a few regions while the demand was dispersed all over the country.

However, such large increases in coal production could not be achieved using the current level of technology. The Meeting felt that in the countries of the region full automation was not likely to be attempted but a gradual and steady increase in the mechanization level in coal-mining was expected. It was noted that even with the general increase in mechanized coal mines, small mines would continue to be operated using traditionally labour-intensive technologies, mostly with a view to supporting local economic and employment interests. The conversion of coal into liquid fuels was not a bright possibility with the price assumption made. However, by the year 2000, the ongoing work in several countries in the conversion of coal to methanol to be used as a fuel blend or a feedstock to chemical industries might be fruitful. That might result in

some coal being used for methanol conversion beyond the year 2000. In the utilization of coal for power generation, increased investment in fluidized bed combustion would take place which would enable the use of the low-grade coal abundantly available in the region.

The Meeting recognized that the production of such large quantities of coal would cause environmental degradation in coal-mining areas and the utilization of coal, particularly near urban and suburban centres, would cause air pollution. Adequate care should be taken to provide for pollution-abatement measures in all coal production and utilization projects.

Hydrocarbons

The Meeting felt that while considering the potential for the production and use of natural gas, the possibilities regarding oil production should be examined simultaneously. In the oil-exporting countries of the region, there was likely to be more emphasis on gas utilization for domestic uses. In countries such as India, which produced oil to meet its oil needs only partially, there would be increased emphasis on the use of natural gas. In the case of China, it was indicated that the natural gas availability even in the future would be somewhat marginal and would be used mostly for petrochemical and fertilizer production. As an urban fuel, it might be used in the transitional stage. China would reach a production level of 200 million tons of oil by the year 2000, but gas utilization was likely to remain limited.

The Meeting noted that even with the strong emphasis on energy conservation and decrease in oil consumption, China might become an importer of oil unless there were more major discoveries of oil fields beyond the year 2000. India and Thailand were likely to combine the use of natural gas for petrochemical/fertilizer production and power generation. Natural gas would be subjected to separation and the appropriate fractions used for different uses. In Indonesia, which was a major producer of oil and gas in the region, owing to possible market saturation for LNG, there was likely to be greater emphasis on the domestic utilization of natural gas. In other oil/gas producing countries, there would also be increased use of natural gas for domestic needs.

Large countries such as India and Indonesia were likely to plan for gas grids for the distribution and utilization of natural gas within the country. There was also a possibility of subregional natural gas grids in the ASEAN area.

The Meeting, however, emphasized the fact that the oil/gas production plans of the region would depend on a number of factors. The recent reduction in the price of oil had led to a slowing down of exploration efforts. The reserve/production ratio of oil and gas had declined in the past five years. Major discovery of oil fields (like Bombay High in India) had not been repeated in the region. The financing of oil exploration activities, either through multinational financing agencies or through production-sharing, should be encouraged. Even though the cost of oil exploration services had been declining in the past two years, not many countries of the region were in a position to take advantage of that situation owing to financial constraints. Considering those factors, the Meeting felt that the increased production and utilization of natural gas in the countries of the region would depend on the ability of the countries to find the resources required for the oil sector on economically and politically acceptable contract terms.

Electricity

The Meeting noted the review of the prospects for power generation presented in the secretariat paper. It was argued that the region was emerging as the one with the fastest rate of growth of electricity consumption, and the required net additions to power generation capacity were the highest in any region of the world. Such a large programme, with a growth rate ranging between 8 and 12 per cent per year, would mean the addition of nearly 7,000-8,000 MW of power each year. That would call for very large investments. Except in China and India, a major part of the investment would be required in foreign exchange. The electricity sector had been traditionally receiving assistance from multilateral and bilateral aid agencies. Unless the flow of funds from those agencies continued and increased, there was danger of the power development programme in the region being severely curtailed. Furthermore, the pricing of electricity in most of the countries had been based on non-commercial considerations. That had led to power-generating agencies not covering the cost fully on commercial terms. That situation had eroded the internal fund-generating capacity of electricity-generating agencies, and had also weakened their capability to raise funds from external financing agencies. The Meeting was therefore of the view that unless administrative and financial reorganization was effected in the electricity-generation agencies, new power plants for the region would not materialize to the required extent.

The Meeting suggested that countries in the region should explore all possibilities of raising funds for power generation, besides aid on bilateral and multinational bases. The suppliers' credit and "build-operate-transfer" offers should also be explored fully. The surplus capacity in the electrical power generation equipment industry should provide an opportunity to negotiate for rapid additions to installed capacity in the countries of the region.

In the operating stage of the power systems, there were a number of economies to be effected. Transmission and distribution losses were very heavy in many countries of the region, while labour employed per unit of installed capacity was also high. There was therefore a need to improve the operational efficiency of electricity agencies.

The Meeting held the view that despite the environmental issues and the unavoidably long lead time, hydro-electric power generation should receive greater attention. In most of the countries, hydropower exploitation had not been as vigorous as it could be. Mini-hydropower generation potential existed in most of the countries of the region and that should also be exploited.

In some countries of the region there was a potential for geothermal power generation on a medium and large scale. Increased emphasis should be placed on the early utilization of that resources.

The Meeting noticed that there were specific programmes for additions of nuclear power generation in Japan and in China, India, Pakistan and the Republic of Korea. As many of the remaining countries in the region had not worked out detailed plans for improving their industrial capability and for producing nuclear power experts and technicians, it was unlikely that there would be nuclear power generation in any of those remaining countries by the year 2000. The Meeting stressed that in countries opting for nuclear power generation in a long-term perspective, efforts should be made to improve the industrial support capability and to build up trained manpower as basic prerequisites.

In China, there were activities for harnessing nuclear energy for uses other than power generation. The possibilities of using nuclear energy to provide low- and medium-temperature heat for district heating and industrial applications were being explored.

New and renewable sources of energy

The Meeting noted the potential for new and renewable sources of energy applications described in the secretariat draft paper.

The Meeting strongly felt that in future there should be greater emphasis on selecting cost-effective technologies which could be applied over large rural areas. Sophisticated hardware which involved trained manpower for operation and maintenance when installed in rural areas could only provide an opportunity to test the equipment with a view to improving their design parameters. The cost of installing and maintaining such equipment was prohibitive for its wider application.

The Meeting identified the wide dissemination of the improved cooking stoves using traditional fuel as the most cost-effective and operationally manageable innovation in the sector of new and renewable sources of energy. For the programme in that area it should be accepted that location-specific characteristics of the stoves should be taken into account; the importance of users' acceptance of the designs was also emphasized. The recent work of the biomass, solar and wind energy network (BSW) had indicated that some very specific models of improved stoves (such as the improved Thai bucket stoves) had found acceptance in specific areas in other countries. It would be useful to arrange for exchange of information on a wide scale so that the preferred design could be identified by different localities. Apart form that, dissemination by other means would have to be the responsibility of the countries concerned. Large programmes for distribution of cooking stoves in China and India could provide the knowledge and insights for designing similar programmes in other countries.

The Meeting felt that the major programme on new and renewable sources of energy which would help not only in energy development but also in rural development as a whole would be the use of biomass (especially agricultural residues) for energy production, including electricity generation. The programmes of agricultural development would lead to increases in agricultural residues. This could be bio-degraded to produce liquid fuels, or used to produce gas or electricity. The appropriate conversion technology had to be selected, with reference to location-specific costs and benefits. Information on techniques of analysing the cost benefits designed recently by BSW could be disseminated widely. Furthermore, specific programmes for utilizing agricultural by-products from agro-industries (such as rice mills and oil mills) to produce electricity for the use of the industry should be pursued. There were a few working models of such units within the region.

The Meeting further recommended that to stimulate private investment in such ventures in new and renewable sources of energy, the legal barrier to the production of electricity by the private sector should be modified in accordance with country requirements. Incentive prices offered to such ventures by the power-generating agencies would stimulate private endeavour in that direction. Even an advanced country like the United States of America had pursued that path, and some countries in the region, such as India, had already announced policy procedures to that effect.

Photovoltaics (PV) technology was not likely to have wide applications for rural energy supplies, but PV appliances would prove to be cost-effective in specific locations and for some special purposes. It was necessary to adopt such techniques with the full knowledge of the requirements and benefits. Large countries with remote areas not connected by the national electricity grids could take the initiative in standardizing peripheral equipment, and in arranging for large-scale local production. That would bring down the cost considerably and help in modernizing (by energy supplies) in some of the most neglected areas of some countries by PV power supply.

The Meeting emphasized the need for integrated rural energy planning and implementation skill as a prerequisite for the wider use of new and renewable resources. Rural energy planning would require skills in assessing the rural energy need, in evaluating new and renewable sources of energy supply options and in configuring the optimum sup-

ply system. Efforts in organizing specialized training for rural energy planners should be intensified. That would provide an opportunity for regional co-operation as there were successful models of exploiting different kinds of new and renewable sources of energy in different countries. The current efforts in that respect for improving the planning and implementation skill should be intensified.

(ii) Consideration of issues on human resources development

(Item 5 of the agenda)

The Meeting had before it secretariat document NR/MSEP/2, "Human resources development", and two other documents on human resources development for energy management presented by the experts from India and Thailand.

The secretariat paper pointed out that the lack of adequately trained manpower for implementing national energy development programmes and projects effectively was a constraint of key significance in energy sector programming. Human resource development had therefore become an important component of the national economic policy of both developed and developing countries. Energy sector manpower needs in the least developed countries and in the island countries were more acute than in other developing countries of the region. A logical approach, suggested in the paper, to practical implementation of training programmes lay in need assessment and in identifying issues relevant to needs. Types and methods of training depended on the requirements. The coverage of manpower assessment should be decided in three directions: by technologies, by economic activities and by manpower categories. Technical co-operation among developing countries (TCDC) and regional/interregional co-operation in energy manpower training might play a significant role in supplementing national efforts. The paper envisaged the TCDC working group mechanism as self-sustaining inter-country co-operation. The concept of networks was also suggested as a suitable approach towards better regional and international co-operation.

The paper presented by the expert from India argued that the immediate concern in the developing countries should be to identify the areas in energy sector management where the availability of adequate skilled manpower proved to be a constraint to energy sector development. The paper then listed certain broad areas where such constraints were noticed and suggested the nature of the training programmes to be undertaken.

Solving the energy issue was an evolving process which required special skills of energy personnel, ranging from policy formulation to implementation. Effective energy planning was believed to be the result of having capable trained personnel and accumulated knowledge. In the implementation stage, technical personnel with an academic background in related scientific and technical fields was needed. In the area of energy policy and planning, the requirement of trained manpower was not so much in quantity as in quality. In the area of energy development and project implementation, the requirement of skilled manpower was wide covering activities such as energy supply development, energy management and conservation, and development of new and renewable sources of energy. Those were some of the points that were elaborated in the paper presented by the expert from Thailand. The paper concluded by emphasizing the role of regional/national institutes linked in human resources development.

The Meeting commended all the papers presented, which had provided useful information on the issue of human resources development as it concerned energy. After in-depth and careful consideration of the papers and other information provided, the Meeting made the following conclusions and recommendations.

Conclusions and recommendations concerning human resources development

The Meeting recognized that the rise and fall in the price of oil in the international market had disturbed the efforts of the developing countries towards self-reliance in the energy sector. The current sense of complacency was partly due to lack of adequate awareness on the part of the general public of the long-term global issues in the energy sector. There was a need to prepare a whole generation of humanity to appreciate energy problems in a long-term perspective and to act rationally in the interests of inter-generational equity. While traditional emphasis on education and training in the technical area should continue, there should be increased efforts to educate the general public on energy issues; youth and voluntary agencies could be involved in such educational efforts. Even in the normal efforts towards upgrading or creating the requisite skills for energy sector management, there should be more emphasis on understanding the energy sector activities as a total system and on the managerial aspects of energy production and distribution than on the technical aspects.

While most of the training could be carried out with national resources, international co-operation was required to provide opportunities for exchange of experience among top management personnel and policy makers and to develop the curricula and build institutional resources. The training of teachers was also an area for international co-operation.

The 1990s would be a period of transition from traditional patterns of fuel use to long-term optimal patterns. The human resources development efforts in the 1990s should provide for the efficient management of the transition period, which would alone ensure the long-term effective management of the energy sector in the developing countries.

The Meeting noted that the higher education facilities in the region were adequate and that personnel with a general and technical background like graduate engineers or diploma holders were available to the extent required. But their knowledge and skills were not quite adequate to fulfil the job requirements. The efforts made by the United Nations and other aid agencies had led to the organization of training facilities in the region. The results had not been very satisfactory, however, for the following reasons: (i) training programmes were not organized with adequate care to impart specific skills relevant to the job requirements; (ii) the programmes had not been designed with reference to teaching materials prepared on the basis of country experience and illustrations; (iii) the trained personnel had not been given opportunities to make the best use of the training owing either to inappropriate placement in the organization or to procedural rigidities within the organization.

The Meeting reiterated that human resources development could not be considered as a mere training programme but should be extended to structural changes in the organization and changes in procedures of work and decision-making which would bring about a sense of involvement and commitment on the part of the trained technical and managerial staff.

Appraisal of the energy enterprises mostly in the public sector tended to show poor performance of the organizational aspects, which was reflected further in the poor performance of the personnel.

The Meeting recommended that the very expensive regional training facilities should be made more effective by reorganizing the design of the programme to ensure that the trainees could cope with the limitations imposed by their organizations.

The suggestion was made that the beneficial results achieved by the quality control circles in Japan and other countries of the region in securing commitment through participation by the employees should be extended to other countries in a TCDC context.

The Meeting recognized that the TCDC and regional/interregional co-operation in energy manpower training could play a significant role in supplementing national training efforts.

(iii) Consideration of the medium-term plan, 1990-1995, in energy
(Item 6 of the agenda)

The Meeting had before it a draft prepared by the secretariat on the medium-term plan in energy, 1990-1995.

The Meeting generally agreed with the draft, to which it made some modifications.

(c) Adoption of the report

The Meeting adopted its report on 15 May 1987.

Annex

LIST OF PARTICIPANTS

to

The Meeting of Senior Experts Preparatory to the fourteenth session of the Committee on Natural Resources.

Bangladesh

Manzur Murshed, Deputy Chief, Ministry of Energy and Mineral Resources, Dhaka

China

Lu Yingzhong, Vice Chairman of University Council, Tsinghua University, Beijing

India

T.L. Sankar, Director, Institute of Public Enterprise, Osmania University Campus, Hyderabad

Indonesia

A.J. Surjadi, Director for New Energy Development, Directorate General of Electric Power and New Energy, Ministry of Mines and Energy, Jakarta

Thailand

Charuay Boonyubol, Director, Energy Research and Training Centre, Chulalongkorn University, Bangkok

*   *   *

Asian Institute of Technology (AIT)

N.J.D. Lucas, Chairman, Energy Technology Division, AIT, Bangkok

7. Prospects of the production and use of coal, natural gas and power generation in China
(Synopsis)

China, a developing country with a population over 1 billion, whose energy consumption is enormous in magnitude, is also generously endowed with natural resources. China has to formulate its policy guidelines governing energy production and utilization in view of its national characteristics in terms of energy resources. It has rich coal deposits with large prospective reserves, so for a considerable period of time, China's energy production and consumption composition will remain coal-oriented. Owing to inadequate exploration, the production of natural gas is still at a low level. China's power industry witnessed a rapid development with thermal power plants as the major contributors. Coal is the major fuel used in power generation.

According to the recent estimate, China's total coal deposit amounts to 4 trillion tons with the identified geological reserve as much as 800 billion tons, ranking third in the world. China also has a rich deposit of natural gas resources based on a geological analysis. The exploration of natural gas, however, remains insufficient at present. The available hydropower energy resources in China amount to 380 million kilowatts. But China's energy resources feature uneven geographical distribution, with 60 per cent of coal centred in Shanxi Province and the Inner Mongolia Autonomous Region in north-west part of the country, 68 per cent of hydro-energy resources concentrate in the south-west region, whereas the densely-populated and industrially-developed parts of the country are east China, mid-south China and the north-east. This situation presents a remarkable challenge to development, transport and distribution of energy resources.

Coal output increased from 32 million tons in 1949 to 872 million tons in 1985, at an average annual growth rate of 9.6 per cent. During the same period, the output of natural gas increased from 7 million cubic metres to 12.93 billion cubic metres at an average annual growth rate of 23.2 per cent; electric power output from 4.3 TWh to 410.7 TWh at an average annual growth rate of 13.5 per cent, of which, the share of hydropower output also increased from 7 TWh to 92.3 TWh at an average annual rate of 14.5 per cent. In 1985, the total output of conventional commercial energy resources (coal, oil, natural gas, hydropower) amounted to 855 million tons of coal equivalent.

China's energy consumption rate is low. In 1985, per capita energy consumption was 731 kg of coal equivalent, about one third of the world's average level. Even in the year 2000, China's per capita energy consumption is expected to be 1 ton of coal equivalent, still at a low level.

China's coal output was mainly used to meet domestic needs. For example, in 1985 annual coal export was merely 7.77 million tons. Of the domestic consumption, 65.6 per cent went to industry, about 24.9 per cent to households, 4.6 per cent to agriculture, 2.8 per cent to transport, 1 per cent to commerce, and 0.7 per cent to construction industry. To meet the needs of the developing economy, China's coal output is envisaged to be 1 billion tons in 1990, and 1.2 to 1.4 billion tons in 2000. This requires strengthening of the coal-mining industry through renovation and expansion of the existing coal mines and exploration of potentials of the old mines.

In China, all natural gas is put to domestic use, of which 96 per cent goes to industry, and 4 per cent to meet household needs. At present, gas supply is insufficient. It is projected that output of natural gas will double the present production by 2000.

In 1985, 91.6 per cent of China's electric power output was utilized by the production sectors such as industry, agriculture and construction; the share of electricity used in non-material production sectors such as households was 8.4 per cent. Though considerable progress has been achieved in China's power industry, it is still unable to meet the needs of the development of the national economy. Forty per cent of the rural population still do not have access to electricity. The current rapid increase in power consumption in both rural and urban areas has further aggravated the power shortage, which has adversely affected development and the living standard of the people. It is envisaged that in 1990 power generation output throughout the country should reach 550 TWh, and in 2000, it should be 1,000 to 1,200 TWh, 3 to 4 times that of the 1980 level. Priority is given to developing thermal power plants, which will be built at the site of coal-mining facilities and the adjacent locations, such as the large coal-mining complex and open-air coal mines. In coastal areas, power generation plants will be installed at the major ports and other sites where substantial power consumption is anticipated. Hydropower stations will be built in the upper reaches of Huang He basin, the upper and middle reaches of the Yangtse River and the Hong Shui He basin. It is projected that in 2000, the hydropower generating capacity will be increased to 70 million to 80 million kilowatts, with an output of 200 to 250 TWh. Nuclear power development has just made its first step; nuclear power plants will be constructed in a systematic and orderly way in localities where prosperity and energy shortage coexist. Rural areas will be encouraged to develop small hydropower stations, wherever hydropower potential is available. Now 100 selected counties will carry out rural electrification pilot projects. The projected hydropower generating capacities will be 20 to 22 million kilowatts, turning out an annual output of 60 to 70 TWh.

Since China's coal and water resources are largely located at the western part of the country, while power users are centred in the more developed coastal areas to the east, high voltage power transmission will be a priority development sector. In the seventh five-year plan period, it is projected that the capacity of the power transmission network will be expanded to 20 million kilowatts. The South China power transmission network will be erected to contribute to the gradual formation of the integrated nationwide power transmission networks in the 1990s.

In China, 800 million people live in the rural areas, where 85 per cent of daily energy needs are met by non-commercial energy resources. Appropriate indigenous energy techniques will have to be developed in view of the specific conditions.

8. Prospects of coal mining development in Indonesia
(Synopsis)

Introduction

Following the first oil crisis in 1973, the Government of the Republic of Indonesia in 1976 issued a policy to increase, as much as possible, the use of coal as an energy alternative for the country, especially for electric power plants and cement factories. The aim is to reduce the increase of domestic oil consumption, and to save oil which has become the most important export commodity for the national income of Indonesia.

In 1978, the Government defined the national policy on the energy sector to include four important subjects: inventorization, diversification, conservation and indexation. Actually, inventorization and exploration for coal resources has been activated since 1974. As a result, plenty of coal deposits have been revealed amounting totally to about 25 billion tons of coal, mainly in Sumatra and Kalimantan.

Based on the government policy on the energy sector and on the potential markets of both domestic as well as for export, it is believed that the future of the coal industries in Indonesia is encouraging. At the present time, both the government and private sectors are conducting many activities for the development of coal mining. Coal

production of the state coal mining companies has been increasing significantly; some new coal mines are to be established soon to meet immediate coal demand for domestic power plants and for export.

(a) Coal reserves

In line with the coal mining development programme, extensive explorations for coal have been carried out since 1974 in South Sumatra and other tertiary coal basins of Indonesia. The investigations are still in progress, conducted both by government agencies as well as private companies.

The total coal reserve is around 25 billion tons consisting of 1,700 million tons of proven, 6,600 million tons of inferred and 16,700 million tons of hypothetical reserves.

The proven reserves are found in the central areas of Sumatra, South Sumatra and eastern parts of Kalimantan; the quantity is sufficient for production of 50 million tons of coal per year for a period of more than 30 years.

Summary of the coal reserves is given in Table 1.

(b) Projected coal productions and demands

Up to the year 2000, coal production in Indonesia is assumed to be much dependent on the two main domestic consumers: electric power plants and cement factories. Besides, substantial demand for coal is expected to come from an oil-field in Riau, central part of Sumatera, for the generation of water steam for secondary recovery. Exports of coal to the neighbouring countries and industrial countries, are also expected to increase gradually.

The establishment of new coal generated power plants offer the most important opportunity for reducing the increase of oil consumption and increasing coal as an alternative source of energy. It is expected that coal will be used also by at least half of the additional power plants in Java in 1990-2000 and by those to be made outside Java.

Table 1.

Summary of coal reserves and resources of Indonesia

(Thousands of tons)

No.	Area	Reserves		Resources	Total
		Proven	Indicated		
1.	North Sumatra	–	1 700 000.0	–	1 700 000.0
2.	Central Sumatra	80 940.5	725 433.5	785 000.0	1 591 374.0
3.	South Sumatra	926 500.0	1 190 000.0	15 904 500.0	18 021 000.0
4.	Eastern Kalimantan	723 025.0	2 980 157.0	–	3 703 182.0
	TOTAL	1 730 465.5	6 595 590.5	16 689 500.0	25 015 556.0

The second largest domestic consumers of coal are cement factories.

(c) Projected coal production

To meet the increasing demand of coal, the production of coal will be escalated soon by both existing as well as new mines.

The open cut and underground mines of Ombilin, West Sumatera, have been rehabilated and expanded to reach a production target of 750,000 tons per year which was successfully achieved in 1985/86. Furthermore, a feasibility study for the expansion of underground mines in the Ombilin II coal field has been made for a production of 600,000 tons per year in the 1990s. In South Sumatera, the Bukit Asam coal mine has opened up a new coal mine at Air Laya for a production of 3.2 million tons. Full production will be made in 1988 to supply electric power plants Suralaya I and II of West Java. Two full feasibility studies have been completed in 1986 for mining two coal deposits in the surroundings of Air Laya, namely Muara Tiga and West Banko. The coal production from Muara Tiga and Banko are to be 2.7 million tons and 5.4 million tons respectively to supply power plants of Suralaya III-VII. In the central parts of Sumatera, an exploration for coal is being conducted by joint technical co-operation between the New Energy Development Agency (NEDO) of the ministry of International Trade and Industry of Japan and the Department of Mines and Energy of Indonesia. The purpose of it is to study the possibility of producing coal for the generation of water steam for secondary recovery of oil; the maximum demand is 7 million tons per year.

In Kalimantan, at the present time, there are 10 companies which have signed contracts with the Government of the Republic of Indonesia for the development of coal mines. Two companies are conducting their feasibility studies, and the others are still in the exploration phase. Besides, smaller local mining companies have started producing coal from many coal fields in Sumatera, Java, and Kalimantan.

To support the programme of coal production, the Department of Mines and Energy has completed a manpower development study on the coal sector. Detailed planning for the implementation of the result of the study will be made soon by a consortium funded by a loan from the World Bank.

Concerning infrastructures, some activities are being carried out by the Government in South and West Sumatera, such as the improvement of the railroad from Tanjung Enim to the coal terminal at Tarahan, rehabilitation of the coal terminal at Kertapati, and construction of new terminals at Tarahan and Teluk Bayur.

An estimated capacity of coal production up to the mid-1990s is shown in table 2.

Table 2.

Estimated capacities of coal production in Indonesia

(Tons)

Coal mines	1989/90	1990/91	1991/92	1992/93	1993/94
Ombilin	1 462 500	1 525 000	1 637 500	1 750 000	1 750 000
Bukit Asam	2 975 000	2 975 000	3 195 000	3 195 000	3 195 000
Muara Tiga	300 000	800 000	2 700 000	2 700 000	2 700 000
Banko	–	750 000	1 750 000	3 000 000	5 400 000
Arutmin	750 000	1 000 000	2 000 000	3 500 000	5 000 000
Utah	250 000	500 000	1 000 000	1 500 000	2 000 000
Kideco	300 000	1 000 000	1 500 000	2 000 000	2 000 000
Kaltim Prima	500 000	2 000 000	4 000 000	6 000 000	6 000 000
Berau	–	–	–	600 000	1 600 000
Adaro	–	250 000	500 000	2 000 000	4 000 000
Multi Harapan Utama	1 000 000	1 000 000	1 000 000	1 000 000	1 000 000
Kitadin	500 000	500 000	500 000	500 000	500 000
Fajar Bumi Sakti	200 000	200 000	200 000	200 000	200 000
Baiduri Enterprise	200 000	200 000	200 000	200 000	200 000
Tanito Harum	500 000	600 000	700 000	800 000	800 000
Bukit Sunur	250 000	250 000	250 000	250 000	250 000
Danau Mas Hitam	250 000	250 000	250 000	250 000	250 000
Cerenti	–	–	–	1 000 000	2 000 000
TOTAL	9 437 500	13 800 000	21 382 500	30 445 000	38 845 000

(d)   Conclusion

In line with the government policy on the diversification of energy sources, it is decided to increase the national production of coal to reduce the increase of domestic oil consumption. Most of the coal will be used by two main consumers, namely electric power plants and cement factories. There are also potentials for coal exports to the neighbouring as well as to the industrial countries.

The proven coal reserve of 1.7 billion tons is sufficient for a production of 50 million tons of coal per year for 30 years, and the total reserve is 25 billion tons.

More investments in the coal chains are expected from private sectors to accelerate the coal mining development.

9. Future trends of electric power in Indonesia
(Synopsis)

(a)   Past and present conditions

In Indonesia the supply of electric power is mainly undertaken by the Government through the State Electricity Corporation (PLN). Besides PLN there are still other parties which are permitted by Government to generate electricity: small private companies and electric co-operatives, particularly in rural areas, generate electric power in relatively small amounts, and captive power plants produce electric power in relatively large anounts for use in industry.

Since the first five-year plan (1969/70-1973/74), the supply of electric power in Indonesia has shown a fast trend of development. Even so, it was still unable to meet the entire demand for electricity in Indonesia.

This is evidenced by the considerable amount of auto-generation (captive) plants installed outside PLN. The electricity production per capita in Indonesia for the year 1986/87 was about 198.8 kWh, of which 115.5 kWh is from PLN, and it is still very low indeed. The total electric energy consumption per capita in Indonesia for the year 1986 was about 171.1 kWh, of which 87.8 kWh was from PLN. And the average total electric energy consumption rose from 23.1 kWh per capita (10.7 kWh is from PLN) in 1967 to 37.6 kWh (17.6 kWh is from PLN) in 1973 and to 128.7 kWh in 1983.

With around 6.97 million electricity consumers in a country of 168.5 million inhabitants only a small percentage of the population has the privilege of enjoying electricity.

Power supply in Indonesia has been characterized by a very heavy reliance on oil. In 1982, almost 83 per cent or 9,832 GWh of PLN power was generated by oil through gas turbines, diesel and steam plants; only 17 per cent of power was generated by hydro and geothermal.

The total generating capacity of Indonesia at the end of fiscal year 1986/87 was 11,817 MW, and for PLN

6,200 MW. Of PLN's total generation of 19,455 GWh, 32.8 per cent came from hydro and geothermal units, 18.1 per cent from coal-fired power plants and the rest from oil-based thermal power plants. The total capacity of the captive power plants is currently about 5,600 MW. Most of the plants are small diesel generating units with some larger gas turbines, steam power plants and hydropower plants, especially used in heavy industries.

Captive power plant installations are primarily in the mining and industrial sectors, and their electricity output is considerably larger than their aggregate purchases from PLN. It is estimated that major industries at present generate about 75 per cent of their requirements. Prominent among the industries producing power are: textiles, food, chemicals, cement, paper, aluminium, iron and steel. The largest captive power plants are Krakatau Steel in West Java (400 MW), Hydropower Plant Asahan in North Sumatra (603 MW) and Larona in South-east Sulawesi (165 MW). In addition, isolated industries, particularly in Kalimantan and Irian Jaya depend upon diesel generating sets.

The rural areas in Indonesia account for about 80 per cent of the population, scattered in more than 61,000 villages throughout Indonesia. The number of electrified villages is around 21.5 per cent.

The bodies/institutions involved in rural electrification activities in Indonesia are PLN, through its Sub-Directorate for Rural Electrification, the Department of Co-operatives and the Provincial Governments.

The manner in which PLN implements the rural electrification scheme is that the prospective villages having access to be connected to the existing grid are given priority provided the villages are economically feasible. For the remote villages, priority was given to utilize the minihydro potential and other renewable energy resources. The construction of small diesel plants was chosen in areas where the above-mentioned alternatives are not available.

(b)   Electric power sector development

The first step in power system planning is to forecast the power demand. For long-range planning, PLN estimates the demand for electricity up to the year 2000. A number of studies have been conducted formulating the power demand for Indonesia. These studies reveal clearly that as a result of the inadequate power supply during the past several years the present growth in power supply is still lagging behind the total actual power demand. This situation is described as a suppressed condition. The power demand for Indonesia until the year 2000 has been projected, as shown in figure I.

Based on the policies of the power sector development and projection of electricity demand, a power system development programme is laid down as shown in table 1. Electric energy production and primary energy needed to meet the demand are shown in table 2 and table 3.

Table 1.

POWER PLANTS INSTALLED CAPACITY DEVELOPMENT
1986/87-2000/01

MW

	1986/87	1987/88	1988/89	1989/90	1990/91	1991/92	1992/93	1993/94	1994/95	1995/96	1996/97	1997/98	1998/99	1999/00	2000/01
1. Hydro p.p	1 240.3	1 411.6	2 033.6	2 169.6	2 192.6	2 319.2	2 330.4	3 082.0	3 094.8	3 652.0	3 696.5	3 757.5	4 090.7	4 949.7	5 251.7
2. Steam p.p	2 486.3	2 520.6	3 385.0	3 710.0	3 840.0	3 840.0	4 005.0	4 505.0	6 177.0	7 757.0	8 932.0	10 132.0	11 747.0	12 922.0	14 747.0
– Residual Oil	1 686.3	1 655.0	2 055.0	1 825.0	1 555.0	1 355.0	1 355.0	1 355.0	1 355.0	1 355.0	1 330.0	1 330.0	1 330.0	1 330.0	1 330.0
– Coal	800.0	865.0	1 330.0	1 730.0	1 730.0	1 730.0	1 895.0	2 395.0	4 045.0	5 475.0	6 675.0	7 875.0	9 140.0	10 340.0	12 165.0
– Natural Gas	0.0	0.0	0.0	155.0	555.0	755.0	755.0	755.0	755.0	905.0	905.0	905.0	1 255.0	1 230.0	1 230.0
– Peat	0.0	0.0	0.0	0.0	0.0	0.0	0.0	0.0	22.0	22.0	22.0	22.0	22.0	22.0	22.0
3. Gas Turbine p.p	1 116.7	924.0	904.0	889.0	889.0	994.0	994.0	1 094.0	1 009.0	1 264.0	1 130.0	990.0	1 160.0	1 660.0	1 960.0
– Distillate Oil	1 002.9	733.0	683.0	668.0	668.0	788.0	788.0	888.0	803.0	1 058.0	1 018.0	908.0	1 118.0	1 618.0	1 918.0
– Natural Gas	113.8	191.0	221.0	221.0	221.0	206.0	206.0	206.0	206.0	206.0	112.0	82.0	42.0	42.0	42.0
4. Combined Cycle p.p (Natural Gas)	0.0	0.0	0.0	0.0	0.0	300.0	600.0	600.0	600.0	600.0	900.0	1 200.0	1 500.0	1 500.0	1 500.0
5. Diesel p.p	1 326.2	1 419.0	1 578.0	1 560.0	1 572.0	1 598.0	1 567.0	1 540.0	1 519.0	1 517.0	1 472.0	1 412.0	1 403.0	1 399.0	1 396.0
6. Geothermal	30.0	140.0	140.0	140.0	140.0	250.0	360.0	360.0	360.0	365.0	365.0	395.0	395.0	395.0	395.0
7. Total	6 199.5	6 414.6	8 040.6	8 468.6	8 633.6	9 301.2	9 856.4	11 181.0	12 759.8	15 155.0	16 495.5	17 886.5	20 295.7	22 825.7	25 249.7
– Oil Based %	64.8	59.3	53.7	47.9	44.0	40.2	37.6	33.8	28.8	25.9	23.8	20.4	19.0	19.0	18.4
– Non Oil Based %	35.2	40.7	46.3	52.1	56.0	59.8	62.4	66.2	71.2	74.1	76.2	79.6	81.0	81.0	81.6

Table 2.

ENERGY SALES, PEAK LOAD AND PRODUCTION
1987/88-2000/01

	1987/88	1988/89	1989/90	1990/91	1991/92	1992/93	1993/94	1994/95	1995/96	1996/97	1997/98	1998/99	1999/00	2000/01
1. Sales (GWh)	17 230.9	19 930.4	23 046.0	26 641.0	30 784.2	35 543.0	41 005.2	46 857.2	53 090.6	59 841.2	67 059.4	74 791.9	83 326.9	92 787.3
Growth rate (%)	16.5	15.7	15.6	15.6	15.6	15.5	15.4	14.3	13.3	12.7	12.1	11.5	11.4	11.4
2. Peak Load (MW)	3 831.7	4 470.9	5 126.5	5 871.6	6 709.0	7 636.6	8 735.2	9 909.4	11 134.6	12 421.8	13 837.0	15 339.3	16 973.6	18 776.2
3. Production (GWh)	22 329.3	25 652.5	29 439.2	33 767.6	38 778.5	44 200.4	50 652.5	57 546.7	64 754.0	72 573.2	80 881.3	89 712.1	99 288.4	110 475.2
– Hydro p.p	4 874.6	6 239.1	7 181.5	7 248.5	7 504.1	7 787.5	9 292.0	9 418.1	11 924.4	12 469.2	12 828.7	13 755.0	15 068.8	16 494.8
– Steam pp	12 453.4	14 077.5	15 921.8	20 253.6	22 509.6	23 810.1	27 243.8	35 413.1	40 560.1	46 197.5	52 082.4	58 590.7	65 135.3	75 338.8
Residual Oil	7 301.0	8 797.1	6 399.4	7 758.4	9 019.2	8 510.3	8 176.6	7 669.4	6 364.8	5 274.1	4 914.9	4 634.7	4 774.4	4 946.4
Coal	5 152.4	5 280.4	9 052.4	9 875.8	9 475.0	10 876.6	14 522.2	23 315.7	30 202.9	36 994.5	43 450.4	49 872.0	56 535.7	66 434.7
Natural Gas	0	0	470.0	2 619.4	4 015.4	4 423.2	4 545.0	4 278.0	3 842.4	3 778.9	3 567.1	3 934.0	3 675.2	3 806.9
Peat	0	0	0	0	0	0	0	150.0	150.0	150.0	150.0	150.0	150.0	150.0
– Gas Turbine p.p	1 226.0	1 216.0	1 148.7	1 234.5	1 607.2	1 715.8	3 165.3	1 641.6	1 244.7	1 078.5	1 041.6	1 073.2	1 432.9	1 620.8
Distillate Oil	871.7	839.0	847.4	909.7	1 053.9	1 103.8	2 467.7	1 088.1	974.0	925.5	893.6	1 004.2	1 384.9	1 558.8
Natural Gas	354.3	377.0	301.3	324.8	553.3	612.0	697.6	553.5	270.7	153.0	148.0	69.0	48.0	62.0
– Combined Cycle p.p (Natural Gas)	0	0	0	0	854.1	3 777.4	3 777.4	3 777.4	3 777.4	5 668.1	7 556.7	9 445.2	9 445.3	9 445.3
– Diesel p.p	2 789.6	3 134.3	4 201.7	4 045.5	4 542.8	4 574.7	4 639.0	4 761.2	4 673.4	4 589.8	4 724.6	4 189.2	5 534.2	4 896.4
Distillate Oil	2 653.6	2 933.3	3 944.7	3 632.5	4 171.8	4 144.7	4 190.0	4 425.2	4 303.4	4 221.8	4 351.6	3 860.2	5 171.2	4 664.4
Residual Oil	136.0	201.0	257.0	413.0	371.0	430.0	449.0	336.0	370.0	368.0	373.0	329.0	363.0	232.0
– Geothermal p.p	985.7	985.6	985.5	985.5	1 760.7	2 534.9	2 535.0	2 535.3	2 570.0	2 570.1	2 647.3	2 658.8	2 671.9	2 679.9
– * Fuel Oil based (%)	49.0	49.8	38.9	37.7	37.7	32.1	30.2	23.7	18.6	14.9	13.0	11.0	11.8	10.3
– * Non Fuel Oil (%) Based	51.0	50.2	61.1	62.3	62.3	67.9	69.8	76.3	81.4	85.1	87.0	89.0	88.2	89.7

Table 3.

PRIMARY ENERGY NEEDED FOR ELECTRIC POWER GENERATION
1987/88-2000/01

10^3 BOE

Type of Energy	1987/88	1988/89	1989/90	1990/91	1991/92	1992/93	1993/94	1994/95	1995/96	1996/97	1997/98	1998/99	1999/00	2000/01
1. Fuel Oil														
– Residual	13 119.1 (31.4)	16 974.1 (34.5)	12 449.4 (22.5)	14 003.6 (22.7)	15 818.0 (22.1)	15 314.0 (19.2)	14 720.0 (16.9)	13 865.7 (13.9)	11 777.9 (10.6)	10 058.6 (8.2)	9 627.0 (7.1)	9 195.5 (6.1)	9 342.3 (5.6)	9 490.3 (5.2)
– Distillate	7 953.9 (19.0)	8 532.7 (17.3)	10 543.1 (19.0)	10 119.3 (16.4)	11 547.0 (16.1)	11 557.5 (14.5)	11 509.8 (13.2)	12 004.8 (12.0)	11 576.5 (10.4)	11 188.0 (9.1)	11 334.5 (8.3)	10 689.6 (7.2)	14 412.9 (8.7)	13 915.8 (7.6)
Sub Total	21 073.0 (50.4)	25 506.8 (51.8)	22 992.5 (41.5)	24 122.9 (39.1)	27 365.0 (38.2)	26 871.5 (33.7)	26 229.8 (30.1)	25 870.5 (25.9)	23 354.4 (21.0)	21 246.6 (17.3)	20 961.5 (15.4)	19 885.1 (13.3)	23 755.2 (14.3)	23 406.1 (12.8)
2. Non Fuel Oil														
– Natural Gas	1 140.1 (2.8)	1 210.3 (2.5)	1 876.8 (3.4)	6 103.9 (9.9)	11 032.7 (15.4)	15 084.4 (18.9)	14 768.7 (16.9)	14 014.5 (14.0)	12 593.7 (11.3)	14 821.3 (12.0)	17 154.1 (12.6)	20 416.6 (13.7)	19 820.2 (11.9)	20 171.0 (11.0)
– Coal/Peat	7 862.0 (18.8)	8 094.3 (16.4)	14 225.7 (25.7)	15 061.0 (24.4)	14 683.2 (20.5)	17 101.0 (21.5)	22 673.6 (26.0)	36 217.4 (36.2)	46 596.9 (41.8)	57 121.1 (46.4)	67 011.7 (49.3)	76 651.6 (51.2)	87 010.3 (52.5)	101 716.1 (55.4)
– Hydro potential	9 727.8 (23.3)	12 450.7 (25.3)	14 331.4 (25.9)	14 465.1 (23.4)	14 975.2 (21.0)	15 540.7 (19.5)	18 543.1 (21.2)	18 794.8 (18.8)	23 796.3 (21.3)	24 883.5 (20.2)	25 601.0 (18.8)	27 449.5 (18.3)	30 071.3 (18.1)	32 917.0 (17.9)
– Geothermal	1 967.1 (4.7)	1 966.9 (4.0)	1 966.7 (3.5)	1 966.7 (3.2)	3 513.7 (4.9)	5 058.6 (6.4)	5 058.9 (5.8)	5 059.4 (5.1)	5 128.7 (4.6)	5 128.9 (4.1)	5 283.0 (3.9)	5 305.9 (3.5)	5 332.0 (3.2)	5 348.0 (2.9)
Sub Total	20 697.0 (49.7)	23 722.2 (48.2)	32 400.6 (58.5)	37 596.7 (60.9)	44 204.8 (61.8)	52 784.7 (66.3)	61 044.3 (69.9)	74 086.1 (74.0)	88 115.6 (79.0)	101 954.8 (82.7)	115 049.8 (84.6)	129 823.6 (86.7)	142 233.8 (85.7)	160 152.1 (87.2)
3. Total	41 770.0 (100.0)	49 229.0 (100.0)	55 393.1 (100.0)	61 719.6 (100.0)	71 569.8 (100.0)	79 656.2 (100.0)	87 274.1 (100.0)	99 956.6 (100.0)	111 470.0 (100.0)	123 201.4 (100.0)	135 901.3 (100.0)	149 708.7 (100.0)	165 989.0 (100.0)	183 558.2 (100.0)

Note: Figure in ( ) is in % to the total.

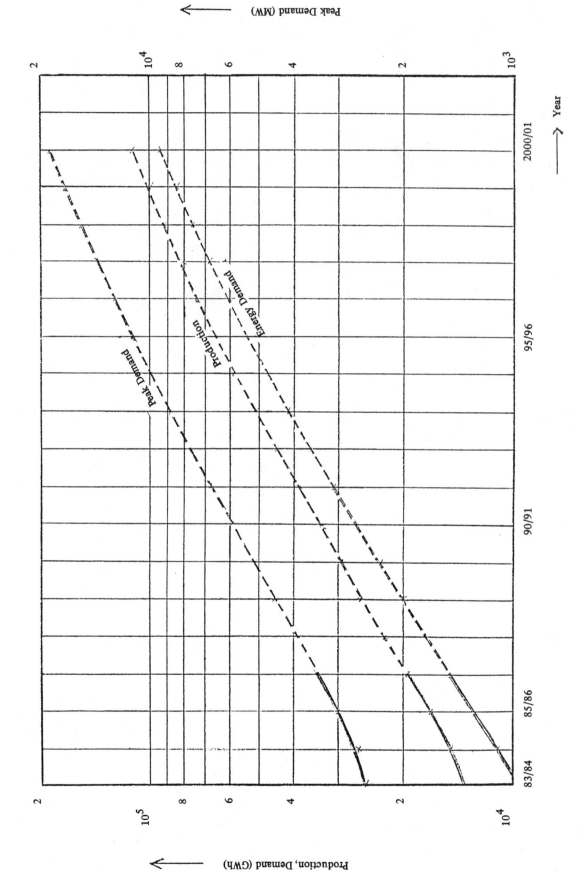

Figure 1. Energy and Power Demand

Development stages will always be in line with the general policies in the energy sector, among others through the diversification of energy resources, such as utilization of hydro potential, geothermal resources and other non-oil fuels such as coal and natural gas. Further increase of thermal efficiency could be achieved through utilization of waste heat from the thermal power plants (combined cycle). The selection among the alternative plant types is based on the least-cost solution.

Indonesia's geography, featuring thousands of islands distributed throughout the archipelago presents specific problems. Java, for example, has an area of only 7 per cent of the total Indonesian territory, but its population is about 60 per cent of Indonesia's total population, and the island's electricity consumption for the current year is 60 per cent of Indonesia's consumption.

Furthermore, Java has limited energy resources with which to supply electric power. The situation outside the island of Java shows quite a different picture. The network system existing on these islands is, in general, of an isolated nature, and their peak loads vary widely. Therefore, the network system development outside Java should be adjusted to the regional development plan.

Development of the power sector in the fifth five-year plan will be directed toward extending power supply to the public in the urban and rural areas, in sufficient quantity and quality, and at a reasonable price level reachable by the majority of the public. The extension of transmission and distribution networks will be implemented, to enable more people to utilize electricity. To increase the effectiveness and reliability of the system, interconnections of isolated systems, and development of control systems will be continued. The main objective of the rural electrification programme will be to electrify all *swasembada* (self-supporting) villages, *kecamatan* (subregency) townships, and villages in the vicinity of distribution lines, and other potential villages. Whenever generating facilities are required, priority will be given to locally available non-oil resources.

Apart from development of power supply facilities, increase of supply quality is needed; this will be achieved through an integrated control system by constructing more load distribution control centres.

Besides the Government, co-operatives and private sectors will be given the chance to participate in the development of the power sector. Development of manpower know-how and capability of local companies is an important aspect in the national economic development. Capacities of local equipment industry should be gradually increased to enable them to participate strongly in the power development.

In order to meet the growing demand for power and the increase of services to the public, it will be necessary to increase the installed generating capacities by about 3,527 MW. During FYP V, energy production will reach 196,859,500 MWh while sales will reach 157,019,600 MWh, and the number of new consumers will increase to 5,894,000.

The rural electrification programme is expected to connect some 2,500,000 new consumers in 9,100 villages.

10. Hydropower in Indonesia
(Synopsis)

An extensive hydro resource survey has been carried out to develop an overall inventory of potential hydro projects so that Indonesia's large hydropower potential could be more systematically exploited. Several studies had been carried out by Electric Power Research Centre in 1968, Overseas Technical Co-operation Agency (Japan) in 1970, and Indonesian National Committee of the World Energy Conference in 1974. A systematic nation-wide resource survey, known as hydropower potentials study (HPPS), was carried out during 1981-1983 with funds provided by the World Bank. The inventory study identified 1,275 schemes with a hydroelectric potential of about 75,000 MW in power or 401,646 GWh annual energy. As part of water resources development (irrigation, flood control etc.), an additional 40 sites have been identified with a hydropower potential of 648 MW of installed capacity.

Approximately 93 per cent of resources exist in the 4 main islands: Irian Jaya, Sumatra, Kalimantan and Sulawesi. The remaining 5 per cent is in Java and 2 per cent in Balí, Nusa Tenggara and Maluku. Hydropower development is limited by their geographic distribution relative to power demand. The greatest potential, over 35 per cent, lies in Irian Jaya where the demand is less than 1 per cent of total domestic power demand, while Java, which accounts for 80 per cent of the current demand, has only 5 per cent of the total potential.

Development types are grouped fundamentally into reservoir and run-of-river type. Estimated potential is classified into 32,985 MW (71 per cent) or 177,809 GWh (65 per cent) for the reservoir type and 13,239 MW (29 per cent) or 97,329 GWh (35 per cent) for the run-of-river type, on the optimum scale.

Among the identified schemes, the total capacity of schemes over 100 MW capacity each is about 32,946 MW, which is equivalent to 57 per cent of the total capacity of the 1,210 schemes. These large-scale schemes are located in the 4 main islands, especially in Irian Jaya and Kalimantan. More than 75 per cent of the potential hydropower resources in Irian Jaya, Kalimantan, Sulawesi and Sumatra fall in acceptable economic range. The undeveloped potential in Java is approximately 1,650 MW at 120 sites. Out of these, about 60 per cent or 1,000 MW fall in acceptable economic schemes.

Only about 2.9 per cent of hydropower potential in Indonesia has been developed and 1.2 per cent is still under construction and about 2.2 per cent is in the detailed design stage.

11. Human resources development in Indonesia
(Synopsis)

Manpower problems in the energy sector

The problems related to manpower in the energy sector in Indonesia are many. First, there is an imbalance between supply and demand of manpower, e.g., quantitative shortages and qualitative deficiencies in manpower. Second is the diversified nature of the energy sector. There is no uniformity in the organization set-up of the manpower planning units in the energy sector: each enterprise has different modes of establishment due to different historical backgrounds, and the foreign enterprises in operation within the sector coming from various countries with different systems of manpower planning are also different in nature.

The Ministry of Manpower in co-operation with the National Planning Bureau (BAPPENAS) made efforts to co-ordinate all activities related to manpower planning, and established the Committee for National Manpower Planning. Members of the Committee came from various ministries and Government institutions, including the Ministry of Mines and Energy. All these efforts will play a part in reaching the Repelita IV target of increased employment opportunities which has been estimated by BAPPENAS to amount to 9.3 million people or 1.8 million per annum.

In meeting key skill demand, various restrictions need to be faced and overcome. There are five aspects to these restrictions:

(1) The nature of manpower in most developing countries;

(2) Manpower supply: the present state of the academies and technical institutes;

(3) Demand on manpower;

(4) Salary and incentives structure;

(5) Skill upgrading schemes to overcome problems of imbalances.

Manpower forecasting and planning

(i) Pertamina

Manpower forecasting planning in the oil and gas subsector is a complex problem covering all aspects of operational activities from production and exploration, refining, marketing, transport to other supporting activities. In Indonesia manpower planning and forecasting viewed with regard to oil and gas operation activities cannot always be based on the assumption that increased production will at the same time increase manpower demand.

Pertamina and its contractors have no arrangement for long-term manpower forecasting and planning and in the case of increasing activities, Pertamina will have third parties or subcontractors to do the required additional activities.

(ii) Electricity

The manpower long-term forecasting and planning is done by using the trend exponential method and time series analysis using data on kWh per employee (kWh employee) since the fiscal year of 1976/1977. In view of the existing management system, type of activities and the size and addition of PLN field of operation, the application of the above formulation is differentiated between the generation unit, project in PLN unit and in PLN Head Office and it is also differentiated between PLN in Java and outside Java.

(iii) Coal

Manpower forecasting in the coal sector is based on a study made by the Coal Directorate in 1985. In the case of the Ombilin Coal Mining, the projection of manpower requirement was based on the production increase from 750,000 tonnes coal in 1985 to 2,210,000 tonnes coal in 1993.

The coal stock piling capacity at the port is planned to be increased to receive 60,000 tonnes of coal. Based on the increasing activities and the operation of additional equipment, manpower forecasting and planning will be calculated on the following criteria:

(a) The number of manpower who are operating mechanical equipment in the mining operation will be increased on the average from 5 to 20 per cent.

(b) Technical supporting manpower will be increased on the average from 5 to 20 per cent.

(c) Supervisory manpower and shift foremen in mining, washing, maintenance, transport, and distribution will be increased on the average from 5 to 10 per cent.

(d) Management/administration personnel will be increased on average from 0 to 10 per cent.

Conclusions

The successful implementation of the national energy programmes requires the support of an effective operational mechanism for manpower assessment and planning. This mechanism will be able to monitor the contribution of the energy sector to manpower supply in relation to the manpower requirements of the programme, and identify the possibilities manpower mobility between different components of the programmes. It is also required to provide a basis for the planning of energy manpower development in relation to the growth of energy development activities and its related sophistication.

There is no consistency in the classification of occupational titles by the respective energy organization or agencies, while the National Standard Classification of Occupation (KJI) published by the Ministry of Manpower is not yet completed and its use for the energy sector is very limited. This is because organization, job and evaluation analyses as a basis for manpower planning activities have not yet been completed in most organizations. On the other hand, institutionalization and methodologies for use in energy manpower planning have not yet been established and developed. As a consequence there are difficulties in assessing the occupational structure of employment in the energy sector or in collecting and consolidating existing information on energy manpower supply related to major occupations in the energy sector.

With regard to skill upgrading schemes a number of isolated attempts have been made by each organization to identify training needs, but these attempts have been impeded by lack of co-ordination and by the fact that the scope and gradation of its training schemes are also different between organizations. Also this is because training policies in the energy sector have not yet been adequately formulated.

Taking the above conditions into consideration, it is still difficult to outline within a certain time span the appropriate overall training development schemes.

Data and information required for ongoing energy manpower assessment are not always available from the existing organizations or agencies. Also much of it is of little value, for in general it is out of date and covers only a limited time-series period. This is part of the problem in the effort of making an accurate analysis and projection of power in the energy sector.

12. Energy-economy patterns and demand management: the Philippine case
(Synopsis)

(a) Energy-economy interaction

Table 1. highlights the impact of energy on the Philippine economy from the crisis years of the 1970s to the present.

From 1973 to 1986 energy consumption grew at about the same rate as population. However, per capita consumption exhibited a generally decreasing trend over the period. Significant decreases occurred starting from the second oil shock of 1978-1979.

As a result of price-induced conservation efforts and de-emphasis on energy-intensive industries, intensity of energy consumption in the Philippine economy continuously moved towards efficiency from 1.169 barrels of fuel oil equivalent (BFOE) per ₱ 1,000 real GDP (in 1972 prices) in 1974 to 0.965 BFOE in 1982 or an 18 per cent improvement. The increase in energy-to-GDP ratio from 1983

when the economy was generally at a slowdown could be attributed to the faster growth of energy consumption of the less productive sectors of the economy.

The substantial devaluation of the Philippine peso against the United States dollar aggravated the rapid escalation of international crude oil prices in the earlier years and offset most of the benefits from oil price decline during the latter years. Thus, while oil prices softened between 1980 to 1986 in the international market, these continued to go up until 1985 as far as the local currency is concerned.

In spite of soaring prices, the country's oil imports continued to rise until the early 1980s owing to the lack of fuel alternatives at that time. With the oil import bill comprising up to about 40 per cent of export proceeds in 1981, the Philippines has suffered continuous trade deficits since 1974, although more recent trends have shown a remarkable improvement.

At the height of the energy crisis in the 1970s, the country was depending on imported oil for the majority of its energy requirements. Intensive indigenous energy development efforts resulted in the availability of geothermal and domestic oil alternatives, in addition to hydropower, coal, and biomass sources. Combined with energy conservation efforts, indigenous fuels to date account for half of the country's fuel supply, a far cry from the 1973 situation when imported oil made up for 92 per cent of total consumption.

Of the total energy use, power generation accounted for about 40 per cent. Power mix, in turn, has been roughly 30 per cent oil-based. The other major oil user is the transport sector, accounting for about 30 per cent of total consumption.

(b) Demand management

The government's strategies towards proper management of energy consumption have employed both price and non-price mechanisms. Non-price intervention includes: (a) quota impositions in times of emergency; (b) direct government intervention to accelerate development of certain markets for non-oil fuels in support of the oil substitution programme; and (c) institutionalization of energy conservation through public and private co-operation of the media, conservation authorities and energy users. Pricing instruments, on the other hand, include price-setting and the use of fiscal measures to achieve certain economic and social objectives.

(i) Pricing policies

In setting energy prices, government has essentially been guided by the principles of economic efficiency, social equity, and financial viability. It is the careful assessment, balance, and interplay of these considerations that determine the resultant market prices of Philippine energy resources from time to time.

Table 1.
Philippine Energy-Economy Interaction

	1973	1974	1975	1976	1977	1978	1979	1980	1981	1982	1983	1984	1985	1986	Average Annual Growth Rate (%) 1973-1978	1978-1983	1983-1986	1973-1986
Energy Consumption (MMBFOE)	69.79	75.19	79.35	83.88	89.73	94.13	97.42	96.87	93.45	95.57	98.47	93.70	92.07	94.70	6.20	0.90	-1.2	2.4
Population (MM)	40.00	41.11	42.07	43.41	44.58	45.79	47.04	48.32	49.54	50.78	52.06	53.35	54.67	56.02	2.70	2.60	2.5	2.6
Energy per capita (BFOE)	1.74	1.83	1.89	1.93	2.01	2.06	2.07	2.00	1.89	1.88	1.89	1.76	1.68	1.69	3.30	-1.60	-4.5	-0.4
Real GDP (BP, 1972 prices)	61.30	64.30	68.40	73.90	78.50	82.80	88.00	92.60	96.20	99.00	99.90	94.20	90.50	90.60	6.20	3.90	-3.3	3.1
Energy-to-GDP ratio (BFOE/'000P real GDP)	1.14	1.17	1.16	1.14	1.14	1.14	1.11	1.05	0.97	0.97	0.99	0.99	1.02	1.05	-0.02	-2.10	2.00	-0.40
Peso-Dollar Exchange Rate	6.76	6.73	7.25	7.44	7.40	7.37	7.38	7.51	7.90	8.54	11.11	16.68	18.61	20.39	2.00	9.10	23.80	9.70
Crude Prices-$/bbl	3.20	10.70	11.60	12.40	13.30	13.50	30.00	36.00	34.00	29.80	29.00	27.00	13.30	16.90	5.50	24.80	-10.20	7.00
Crude Prices-P/bbl	22.00	72.00	84.00	92.00	98.00	100.00	221.00	270.00	267.00	290.00	331.00	483.00	502.00	270.00	52.40	33.00	1.20	33.10
Oil Import Bill (CIF, MM$)	231.00	681.00	833.00	917.00	1 040.00	1 034.00	1 597.00	2 516.00	2 534.00	2 151.00	2 116.00	1 562.00	1 453.00	797.00	48.00	19.20	-26.10	19.80
Oil Fraction in Import Bill (%)	12.90	19.60	22.10	23.20	24.40	20.10	24.10	30.30	29.90	26.00	26.10	26.00	28.00	16.00				
Oil Fraction of Export Proceeds (%)	11.00	22.90	31.90	30.90	27.80	36.20	30.20	38.40	38.60	36.90	37.50	29.00	31.00	16.00				
Trade Imbalance (MM$)	307.00	-497.00	-1 168.00	-989.00	-531.00	-1 197.00	-1 322.00	-1 740.00	-1 908.00	-2 433.00	-2 149.00	-679.00	-482.00	-202.00				
Oil Fraction of Trade Deficit (%)	0.75	-1.37	-0.71	-0.93	-1.96	-0.86	-1.21	-1.45	-1.33	-0.88	-0.98	-2.30	-3.01	-3.95				
Oil Share to Total Energy (%)	92.00	80.10	81.80	79.60	81.00	80.00	72.40	71.70	71.90	68.40	64.50	56.20	51.30	52.60				
Indigenous Share to Total Energy (%)	8.00	19.90	18.20	20.40	19.00	20.00	27.60	28.30	28.10	31.60	34.50	42.10	44.10	44.00				
Power Share to Total Energy Use (%)	26.00	24.60	25.70	26.30	25.60	29.00	28.50	29.80	32.50	33.70	35.40	37.50	41.70	40.80				

Domestic prices of petroleum products are set by the government, taking into account international oil prices, peso-dollar exchange rate fluctuations, and also revenue generation and industrial viability considerations. In order to minimize the inflationary effect of frequent oil price adjustments, domestic fuel prices have not been reduced totally to reflect recent oil import price levels. The government also established an oil Price Stabilization Fund (OPSF) to stabilize domestic petroleum prices and prevent a "roller coaster" effect in periods of unstable international oil prices and peso-dollar exchange rates. The tax component of petroleum prices consists primarily of import duties on crude oil and ad valorem taxes on petroleum products, the level of which are influenced by social equity considerations.

The setting and regulation of electricity tariffs is a decentralized function distributed among the government entities involved in the power sector. The state-owned power generating company or the National Power Corporation (NPC) sets its own rates and submits them for the approval of the Cabinet. The National Electrification Administration which is tasked to implement a nationwide rural electrification programme, sets the tariff structure of, and monitors its implementation by, electric co-operatives. The Energy Regulatory Board (ERB) approves the rates charged by privately-owned electric utilities. The NPC tariff structure is a two-part tariff consisting of a demand charge per kW and a block energy charge per kWh. The rates and the blocks are different for each category of customer, i.e. utilities, industries and non-utilities. Block energy charges are progressive for the utilities and regressive for the industries. The tariffs do not differentiate between voltage levels. Added to the basic rate are fuel cost (oil and steam) and foreign exchange adjustment clauses.

The tariffs of MERALCO, the biggest distribution utility operating in the metropolis and nearby areas, are also based on the type of consumer, i.e. residential, commercial, industrial, and also do not take the supply voltage level into account. From 1974 to 1981, customers enjoyed full generation and distribution cost subsidy for the first 200 kilowatt-hours of residential consumption and 90 kWhs of commercial consumption. The MERALCO rate structure was reoriented towards eliminating the subsidy burden on the industrial and other high volume users, and beginning in February 1985, a five-year phased-in subsidy reduction programme was implemented which will ultimately reduce the distribution cost subsidy levels for both residential and small commercial customers to 50 kWh by 1990.

The need to effect further policy reforms in the power sector was endorsed by the Cabinet after a series of public consultations in 1986. The Cabinet directed an exhaustive study of the power sector by key agencies in the government, and the following pricing policy reforms were proposed and are now under implementation: (a) restructuring of NPC power rates using the long-run marginal cost approach; (b) continued implementation of MERALCO grid distribution subsidy reduction and further reduction of subsidy on generation cost.

(ii) Non-price demand interventions

The government resorts to more aggressive means of intervention in public consumption when extraneous factors interfere to drastically disrupt the normal supply of energy resources. In its most extreme form, intervention was exercised through physical control of products, such as the rationing of petroleum products (once, at the height of the 1973 oil crisis) as well as load shedding and rotation of power cuts in the electricity subsector. Legislations have also been enacted to prohibit unnecessary or unproductive energy consumption. Normal intervention measures include regulation of importation, exportation, shipping, marketing, distribution, refining and processing, and storage activities of the energy industry.

The more passive yet generally effective policy tools in the management of energy demand include education and information dissemination activities directed at making target individuals aware of national energy problems and government's thrusts in conservation and energy management.

The thrust of the energy conservation programme implemented by the government through its line agency, formerly known as the Bureau of Energy Utilization (BEU), is directed towards the industry, transport and commercial sectors in view of their large shares in the total energy consumption.

The energy conservation programme includes the following activities:

(a) Energy management training courses;

(b) Energy management seminars;

(c) Energy briefings;

(d) Energy conservation publication;

(e) Energy audit assistance;

(f) Industry energy efficiency programme.

Indicators to assess the success of the country's energy management effort since the oil crisis of 1973 show the following results:

(1) The ratio of total primary energy demand to gross domestic product in 1986 as compared with 1973 has declined by 8 per cent;

(2) Gasoline consumption has continuously declined over the past 11 years from 14.6 million barrels in 1974 to 9.1 million barrels in 1986 as a result of the switch of public transport vehicles to the more efficient diesel engines;

(3) There has also been consistent increase in utilization of indigenous energy over the years. The displacement

of imported oil by local energy sources was equivalent to 41.7 million barrels in 1986 alone. Moreover, because of lower oil importation and lower crude prices, the country's oil bill dropped by 68.5 per cent from $2,534 million in 1981 to $797 million in 1986.

13. Prospects for production and utilization of natural gas in Thailand
(Synopsis)

Towards the end of this century total energy demand, forecast to grow annually at 6 per cent, is expected to reach 800,000 barrels per day of oil equivalent. The indigenous resources (oil, gas, hydropower and lignite) is expected to constitute 65 per cent of total energy share. In an attempt to ensure security of supply and meet the nation's own future energy demand, Thailand is, therefore, determined to pursue the objectives set forth in the sixth National Economic and Social Development Plan (1987-1991) with emphasis in the following areas:

(a) Accelerating petroleum exploration and development activities to increase domestic petroleum production in order to reduce dependency on imported oil;

(b) Expansion of the natural gas market;

(c) Increase in the nation's refining capacity;

(d) Diversification of the sources of oil supplies.

In an effort to achieve the goal to reduce the dependency on imported oil to 35 per cent of total energy consumption, the Government of Thailand has expressed commitment to maintaining a positive investment environment to encourage foreign oil companies to continue to participate in exploration and development activities.

A recent estimate puts the figure for total current reserves of offshore/onshore natural gas at approximately 16 tcf.; reserves of onshore crude oil total 47 million barrels (MMBBL) and condensate 207 MMBBL.

In the onshore areas, natural gas accumulations have also been discovered. Additional onshore and offshore oil continue to be discovered.

Present combined production of indigenous oil, condensate and natural gas has increased from an initial 5,000 barrels per day (b/d) in 1981 to approximately 116,000 b/d of oil equivalent in 1987, of which crude oil constitutes about 15,700 b/d, condensate 16,200 b/d and natural gas 500 million standard cubic feet per day (MMSCFD). Currently, the total indigenous petroleum production has been contributing up to 30 per cent of total petroleum share in the country.

The natural gas found in Thailand's offshore and onshore fields is rich in ethane, propane, butane and pentane-plus components. The ethane and propane fraction extracted serve as the primary feedstock for the upstream component of the petrochemical complex which converts them into ethylene and propylene as its products.

The demand for LPG alone has grown very quickly since 1984 and is projected to reach over 800,000 tons in 1990. Preparations are, therefore, in hand to call bids for the construction of a second gas separation plant. In the meantime, preliminary study has begun on the development of the second petrochemical complex in Thailand in an effort to maximize the benefits from locally available hydrocarbons. The complex will be aromatics-based using naptha from local oil refineries, indigenous condensate and NGL as feedstock for the upstream plant with downstream output involving various kinds of petrochemical products.

The natural gas demand is forecast to increase up to 700, 900 and 1,000 MMSCFD by the year 1990, 1995 and 2000 respectively. The natural gas demand for power generation shall be maintained as the base load. Adequate gas supply will be produced from the Nam Phong Field "B" structure as well as UNOCAL's other fields.

C. Strategies required in the accelerated development of new and renewable sources of energy

1. Updated assessment of the contribution of new and renewable sources of energy to regional energy supply*
(E/ESCAP/NR.14/5)
(Synopsis)

Traditional energy supplies are still the crucial and often the only available supplies for rural dwellers which represent 80 per cent of the people in the region. However, experience with efforts over the past one and a half decade has shown that biomass resources are unlikely to compete with conventional supplies in providing expanded rural needs for energy. Thus, satisfying regional energy requirements will require a mix of renewable and non-renewable resources.

The paper recommends that:

(i) Regional countries intensify rural energy planning efforts, with both renewable and non-renewable supplies.

(ii) Research, development and demonstration efforts be directed to ensuring adequate supplies of traditional biomass fuels to rural dwellers, concentrating more on users' needs.

(iii) Technical, economic and social research be conducted to reduce cost and increase users' acceptance of renewable technologies.

---

* Note by the ESCAP secretariat. Full text of the paper has been published in *"New and renewable sources of energy for development"*, a United Nations publication, ST/ESCAP/580, 1988.

2. Co-operative research, development and demonotration achievements and future plans regarding new and renewable sources of energy.*
(E/ESCAP/NR.14/6)

Introduction

In August 1981, there was a great sense of expectation: the United Nations system had taken a significant step towards coping with the energy crisis. Subsequently, programmes and projects were drawn up to accelerate the exploitation of new and renewable sources of energy for a sustainable energy future.

Some results have been achieved: fuel oil has been largely replaced in power generating application, if not by renewable sources of energy then at least by cheaper energy sources with a longer time-span; this is a significant achievement.

The penetration of new and renewable sources of energy, however, was not only limited by shortage of capital: it is even more disturbing that in some cases where capital was in fact available at a level at which a significant contribution could have been made, implementation of new and renewable sources of energy projects related to new technologies constituted no more than 10 per cent of capital completion in the five years 1981-1986: the human resources bottle-necks of these new technologies was seriously underestimated and that is the most significant lesson learned.

(a) New approaches

(i) The working group concept (definition)

While implementing energy (or other development-related) projects, professionals charged with the planning and execution of such activities have found that contact with professionals facing similar problems helps them overcome obstacles more quickly, since other people may have already overcome similar problems elsewhere. This is the central idea behind TCDC (technical co-operation among developing countries).

(ii) Co-operative projects

Based on the above working group concept, and encouraged by project fund limitations, it has been found that co-operative projects, where expertise, equipment, labour and capital may all be provided by different participants, have become more and more the model for successful new and renewable sources of energy projects, where training components are automatically built in to the project design from the very beginning. One such example is the ESCAP technology for development programme, including training and pilot projects in solar photovoltaic electrification for remote communities. This project is funded by the Government of Japan (at the planned level of about half a million dollars for 1988-1989 and a similar level in the current biennium) and executed by the biomass, solar and wind network of the Natural Resources Division of the ESCAP secretariat, with overall administrative guidance from the ESCAP/UNIDO Division of Industry, Human Settlements and Technology. In Asia, it is now supplying equipment for training centres in Indonesia and Pakistan, with tailor-made and well-packaged training materials developed at the Asian Institute of Technology (AIT). The Pacific energy development programme (PEDP), a UNDP-funded, ESCAP-executed programme for the Pacific island countries, is contributing with expertise and a specific training/demonstration project at Palmerston Island (in the Cook Islands) where one island is completely electrified with solar photovoltaics. The Pacific expert involved in this is also involved in training activities in Maldives, in the Indian Ocean, where the problems faced are similar to those faced by Pacific island countries. All the above is carried out under the energy programme of the Natural Resources Division of ESCAP.

(iii) Suggested issues for the 1990s

a. Oil prices

It is generally considered likely that because of the structural changes associated with economic development, there may be a strong upward pressure on oil prices again around the mid-1990s. One issue for continued attention by the United Nations system could be how to anticipate, and thus minimize, the adverse effects of this.

b. By-products

One example is supplied by the sugar industry switching to bagasse fuel, a by-product, for its major fuel requirements. Similar adjustments to utilize waste materials partially have also taken place in the palm-oil industry. Biogas systems could be considered for the utilization of by-products of pig-farming, cattle-raising or rice-milling. Continued emphasis on such utilization of by-products could be a second issue on which the United Nations could focus, especially because such systems require a higher order of management expertise than simple, linear, one-product, "naive" production systems.

c. Rural incomes

Last but not least, the problem of rural poverty is still with us: the bottle-neck is not so much energy as rural income levels. An issue of more cash crops to support rural income levels through some energy products may become viable again in the mid-1990s.

(b) Summary

New and renewable sources of energy will still be very crucial in the supply of energy on a sustainable basis,

---

* Note by the ESCAP secretariat.

particularly for rural areas. Thus, it will be essential for the international community to undertake concerted promotional activities to assist developing countries in their national efforts to manage their energy resources, both conventional and non-conventional. Activities in the next medium-term plan period should therefore cover all sources of energy and in all areas such as planning, development and utilization of energy in the light of the structural changes related to economic development that are expected to result in increased energy requirements, as shown by policy studies reviewed by the recent first Asian Forum on Energy Policy. The continued importance of new and renewable sources of energy for a long-term sustainable energy future is also recognized in the list of activities. In order to avoid possible duplication with other agencies, ESCAP involvement in new and renewable sources of energy will be concentrated on selected matured technological areas such as solar photovoltaic, solar drying and cooling, etc., giving emphasis to training, advisory services and information exchange.

The role of the secretariat will be to facilitate co-ordinated energy planning, development and management through advisory missions, studies, and information exchange on both the technological and socio-economic aspects of energy systems. It is envisaged that such a role will be fulfilled more and more through the technical and administrative support of working groups formed by member country organizations in subject areas such as energy planning, rural energy, coal, gas, electricity, new and renewable energy technologies, etc. It is envisaged that such groups would be formed by the end of 1991. Within this general approach, an overview of achievements regarding new and renewable sources of energy is given in annex I and a concrete co-operative programme outline for the medium term is presented in annex II. The Committee is invited to endorse the approach, and to indicate interest in participation in the co-operative programme.

Annex I

ACHIEVEMENTS IN CO-OPERATIVE RESEARCH, DEVELOPMENT AND DEMONSTRATION

Following the recommendations of the United Nations Conference on New and Renewable Sources of Energy (1981) and the Regional Expert Group Meeting on New and Renewable Sources of Energy (1982), ESCAP has carried out many projects promoting co-operative research, development and demonstration in new and renewable sources of energy in the region through activities implemented by its Energy Resources Section and the regional network on biomass, solar and wind energy as well as the UNDP-funded regional energy development programme (REDP) and Pacific energy development programme (PEDP).

Emphasis has been given to the demonstration of newly-established facilities dealing with biomass, solar energy and small hydropower.

Various types of small and mini hydropower generation systems were demonstrated through co-operation among the participating countries in the regional network for small hydropower (RNSHP) in order to promote a deeper awareness of this technology and highlight the necessary measures for application in each country. This led to cross-country co-operation in setting up small hydro facilities through TCDC, such as the co-operative technology transfer between China and Fiji.

In the field of solar energy applications, there have been many co-operative activities in research, development and demonstration, such as solar drying in Indonesia and Thailand and solar thermal in Thailand, to name a few. These activities have resulted in improving each country's facilities and upgrading the technical capability. Another example is in the area of solar photovoltaic (PV) electricity generation, in which TCDC exchange of PV systems for evaluation is under way following a recommendation of the Regional Expert Seminar on Solar Photovoltaic Technology held at Bangkok in June 1985. Countries advanced in this technology, such as China and India, would provide information on these systems leading to TCDC co-operative programmes with interested countries in the region such as Indonesia, Maldives, the Philippines, Thailand and Viet Nam.

Another co-operative research, development and demonstration project in solar photovoltaic technologies is under way based on the Tokyo Programme on Technology for Development in Asia and the Pacific which was adopted as one of the resolutions by the Commission at its fortieth session. This project consists of two phases of demonstration-cum-seminars/training courses organized in Indonesia and Pakistan. Following an expert group meeting in June 1986, the first programme started in Yokjakarta in mid-January 1987 with a seminar/training course on the evalua-tion, design and implementation of solar photovoltaic systems in developing countries. A similar project will be conducted in Pakistan in a first phase, to be followed by more specialized courses in the second phase in 1988. A component for the Pacific island countries and Maldives, covering roving photovoltaic installers' training courses, has been added to the project.

As for solar thermal technology, workshops/training courses have been organized aimed at promoting regional co-operation. One of the typical examples was a training course on solar hot water systems organized jointly with AIT (Asian Institute of Technology) in 1985. It was agreed in the course that co-operation should focus initially on two areas: (i) adaptation of standards to suit regional conditions for solar collector testing, and (ii) performance evaluation of solar systems. Following up these conclusions, ESCAP is planning to conduct study missions to establish regional co-operation between countries in this field.

Reviewing the above ESCAP activities, it is considered that regional co-operation in research, development and demonstration on new and renewable sources of energy has been achieved in some cases. However, much more remains to be done. In order to accelerate regional co-operation, follow-up action should be taken in each field of such energy sources. For instance, to provide the necessary data to facilitate further application of solar photovoltaic technologies, it is considered essential to conduct co-operative climatic measurement of solar insolation and its local characteristics and to establish the standard testing method of efficiency for solar heat exchanging systems under regional co-operation so that each country can improve its solar thermal facilities based on a common standard.

As for the field of research and development, there is still some room for promoting co-operation among the countries of the region, especially multilateral regional co-operation, in addition to the present ongoing bilateral co-operation. In order to facilitate the smoother transfer of technologies of new and renewable sources of energy and their adaptations, it is necessary for the developing countries of the region to improve their technology levels so that they can catch up with developed countries where new technologies have mainly been developed. Research and development conducted co-operatively among the developing countries should foster improvement of their technological capability.

For the first step of promoting regional co-operation in research and development, it is recommended that a tripartite consultation meeting be convened among regional research institutes (one of which could be considered as being able to act as a regional centre for research and development), potential donor countries and institutes, and interested developing countries, to discuss ways and means for co-operative research and development, and to identify priority themes for research.

The Committee may wish to consider these issues and give the secretariat its views and guidance, especially concerning the concrete proposals embodied in annex II to this document.

Annex II

OUTLINE OF A PROGRAMME OF TECHNOLOGICAL CO-OPERATION AMONG DEVELOPING COUNTRIES TO ACCELERATE THE UTILIZATION OF NEW AND RENEWABLE SOURCES OF ENERGY

Objective

To promote and implement a programme of technological co-operation focusing on exchange of equipment and expertise, and thereby accelerate possible equipment manufacture and utilization of new and renewable sources of energy in the ESCAP region.

Background

The second meeting of ESCAP focal points on new and renewable sources of energy held in September 1986 recommended that projects on technical co-operation, training, demonstration and dissemination of technical information should be emphasized by ESCAP. There has been a great deal of research, development and demonstration of new and renewable sources of energy technologies. Both developed and developing countries, and also aid donors, are now emphasizing in their programmes the application of the knowledge gained and commercialization of technologies. ESCAP is well placed to take a leading role in promoting technical co-operation activities that will lead to increased development, manufacture and use of such sources of energy.

Activities aimed at promoting technical co-operation

A major focus of activities in 1987 is the development of concrete programmes of technical co-operation between developing countries of the region, including developed countries wherever appropriate.

Proposed activities include:

(i) Expert seminar on wind energy utilization for water lifting:

The objectives are given in appendix I and, as can be seen, are directed at a concrete co-operative programme of technological exchanges.

(ii) Expert seminar on biomass combustion technologies:

The objectives are similar to those for (i) above.

(iii) Development of a programme of expertise and equipment exchange on cooking-stove technology:

Based on a mission (and review report) that is taking place from March 1987.

(iv) Regional co-operative programme on research, development and demonstration of solar photovoltaic technology:

Phase II of this project will propose utilization of the computer-based design aids developed in phase I for training, design and specification of systems (see appendix II for further details). The aim will be to maximize local involvement in equipment specifications and develop technical co-operation at government and commercial levels between countries of the region and with selected developed countries. India and China have both offered to provide TCDC services in photovoltaics. In addition to the financial assistance given by Japan, Australia, France and the United States of America have provided experts to assist in the programme; the United States expert has indicated support for a more full-scale involvement, possibly also with equipment support.

(v) Co-operation in the evaluation, testing and development of solar heating technology:

Solar hot water systems are already being manufactured in a number of countries of the region. It has been demonstrated in developed countries that field evaluation of systems performance and laboratory testing can lead to significantly improved performance of future systems through modifications in design.

A project on this topic has been discussed with the French authorities concerned, who are examining possible funding.

(vi) Solar drying and energy conservation in tobacco-curing:

A TCDC meeting to review the latest developments and formulate a programme of exchange is planned at Chiang Mai University in late 1987. The University has undertaken extensive work in the field with substantial aid from developed countries.

Proposed funding of a project on regional technical co-operation in new and renewable sources of energy

The activities outlined above will result in a list of proposals for expertise, equipment exchange and evaluation. Funding will be required to implement these exchanges and it is proposed that a project document be developed and funding sought from aid donors. The objectives of the project would be to implement a programme of regional co-operation in technologies and networking arrangements in new and renewable sources of energy at the expert level under the following headings:

(a) Equipment exchange and manufacture

       Commercial equipment and design information

(b) Exchange of expertise

(c) Equipment testing

       Laboratory and field testing

(d) Development of standards and procedures

(e) Comparison and use of design techniques and models

(f) Comparative analysis of equipment performance

(g) Development of guidelines for selection of equipment

(h) Marketing and manufacturing studies

(i) Socio-economic studies

The project would also have the secondary objective of facilitating interregional co-operation in new and renewable sources of energy.

It is envisaged that in cases where the equipment and expertise come from a developed country, funding would be sought from the country concerned. The funding requested for this project would be used mainly to finance manufacture of equipment, training and exchange of expertise of developing countries in the programme.

The photovoltaics portion of the project (phase II) is expected to cost $US 650,000 (1988 — $500,000, and 1989 — $150,000). It is proposed that an additional $US 450,000 be sought from aid donors to finance the non-PV activities for two years, and cover the cost of a possible programme co-ordinator, and the technological exchange activities arising out of the preparatory activities conducted by the regional network on biomass, solar and wind in the areas of solar thermal and wind energy and biomass combustion.

Conclusion

The implementation of a comprehensive project on co-operation in new and renewable sources of energy would put ESCAP in a central role in the promotion of the use of such energy sources in the region and could attract significant support from developing countries in the form of "follow-on" bilateral projects with ESCAP members.

Appendix I

OBJECTIVES OF THE EXPERT SEMINAR ON WIND ENERGY UTILIZATION FOR WATER LIFTING

1.   Review the status of wind water pumping technology, manufacture, research, development and demonstration and evaluation and test activities in the Asian and Pacific region.

What are regional groups trying to do and why, and what are the problems and achievements.

2.   Examine the scope for wind energy regional co-operation and networking arrangements at the expert level and, within this context, develop a TCDC evaluation programme under the following headings:

Wind pump testing

Laboratory

Field testing and demonstration

Development of standard procedures

Exchange of data

Comparison of design techniques and models

Comparative analysis of performance with diesel-drive pumping system

Development of guidelines for selection of equipment

Marketing and manufacturing studies

Socio-economic studies of wind pumping

3.   Develop a TCDC programme of exchange of expertise and equipment

4.   Facilitate interregional co-operation in wind water pumping

Appendix II

ESCAP REGIONAL CO-OPERATION PROJECT ON RESEARCH, DEVELOPMENT AND DEMONSTRATION OF SOLAR PHOTOVOLTAIC SYSTEMS FOR RURAL AREAS

The objectives of this programme are to:

Make countries of the region aware of the potential of photovoltaic technology to suit a wide range of energy requirements

Provide countries of the region with the skills to evaluate, design and implement photovoltaic systems and make decisions on sourcing of components

Assist countries in designing photovoltaic policies, programmes and projects.

Substantial progress has been achieved, as follows:

Expert Group Meeting on Evaluation, Design and Implementation of Solar Photovoltaic Systems in Developing Countries, Thailand, June 1986

Preparation of training manuals and computer packages on photovoltaic design

Training of Indonesian and Pakistan trainers in the application of computers to photovoltaic systems design and evaluation and also in experiments to develop and evaluate photovoltaic components and systems

Seminar and training course on evaluation, design and implementation of photovoltaic systems, Indonesia, January 1987, with 21 countries represented and 90 participants

Commencement in February 1987 of roving training courses for Pacific countries for technicians on installation and maintenance of photovoltaic systems, in which 300 trainees from 17 countries are being trained

The next (1988) phase of this project will build on the progress made in phase I:

The training materials and computer packages will be revised and supplemented by additional materials, with a view to their widespread dissemination

Case studies for use in future training courses will be undertaken utilizing the computer packages. These will be based on real data from countries of the region and include evaluation design and engineering specifications. Assistance will also be sought to implement the projects which are the subject of the case studies.

Training courses to suit different country needs and different target groups, including:

Course on engineering design of photovoltaic systems

Roving course on photovoltaic utilization for remote communication systems, including optical fibre and satellite systems power requirements

Roving course on pumping, village power and telecommunications

Follow-up course for Pacific technical institutes on installation and maintenance of photovoltaic systems.

All of these courses will have a practical orientation with "hand-on" experience through hardware experiments and use of advanced design aids.

Assistance to countries of the region in developing national photovoltaic policies, programmes and projects and in seeking funding support.

3. Energy research: directions and issues for developing countries*
(Synopsis)

The Energy Research Group was convened by the International Development Research Centre and the United Nations University to survey energy research and to suggest energy research priorities for developing countries. The report is based on three premises: (1) energy research must be related to research on the entire economy and society; (2) energy sources must be studied in the context of demand for them; and (3) energy saving is as important as energy production. The report is based mainly on published information. It does not cover engineering and technological details, does not go into regional and national variations, and excludes areas where confidentiality of information or lack of knowledge makes it difficult to judge research.

The Energy Research Group consists of 11 energy specialists[1] from developing countries. The Group has concerned itself with building research capacity, without which priority-setting in research would be pointless. It has tried to identify promising lines of research from the role of energy in development. It has covered all forms and uses of energy.

---

* The summary of the publication was presented by Mr. Ashok V. Desai, Co-ordination, Energy Research Group.

For full text see the title publication of IDRC and UNU 1986 ISBN-0-88936-479-6.

[1] Messrs. Ashok V. Desai, Djibril Fall, José Goldemberg, José Fernando Isaza, Ali Kettani, Ho Tak Kim, Mohan Munasinghe, Frederick Owino, Amulya Reddy, Carlos E. Suarez, Zhu Yajie

The Group started with the following normative assumptions: the growth of production and consumption is essential for a rise in the living standards of developing countries, but deliberate policies are also necessary to relieve poverty and improve income distribution. This concern for fairness extends to the international sphere. The capacity of developing countries for independent decision-making needs to be increased; this requires the internationalization of a number of functions and activities that they lack or for which they depend on industrial countries. Economic activities should pass the test of environmental soundness, and they should be conducted efficiently, although efficiency must be defined in the local context.

The report comprises 15 chapters dealing with various aspects of energy research areas. These include: research and its environment; demand analysis and management; energy conservation; different fuels; environmental effects; resources etc.

In the research and its environment area, the report says that research needs to be useful as well as sound; it depends on the effective interaction of the "doers", directors, and users of research.

Governments are institutions for resolving conflicts; whatever their philosophy, the central role of energy, the scale of energy investments, and the import costs of, or export earnings from, energy force governments in developing countries into action on energy. Energy policy has to take a forward view of feasible futures, choose among them, and select instruments for achieving national objectives; research can improve decisions at each step. For research to effectively assist policy, however, it is essential that it be done by well-endowed, professional research institutions capable of giving independent advice. The government should be an informed buyer rather than an owner of research.

Technologically sophisticated observation of their environment is within the capacity of many firms in developing countries and needs to be encouraged. Government assistance should be directed to building this capacity, and not to research and development as such, and to creating competitive market structures that induce firms to innovate rather than to generate particular innovations. Research institutions working for small firms should be in close touch with some producers, and should carry innovations to the point where the risks of commercializing them are minimized.

For effectiveness, research institutions need to accumulate experience, diffuse intellectual skills, ensure efficient use of their intellectual assets, and bring together diverse disciplines to bear upon problems. Programme funds to them should be directed toward building intellectual and material assets in specific areas of research; projects should be designed to exploit those assets. Directors of research institutions play a crucial role in co-ordinating researchers and problems, programmes and projects. Training and communications are essential to good quality in research.

4. Efforts to improve the utilization of new and renewable sources of energy in Indonesia
(Synopsis)

(a) Programme implementation

Various projects have been implemented during the past several years to develop new and renewable sources of energy, which covers activities such as training and research, development, and demonstration, as well as efforts toward commercialization of such resources. Activities have been concentrated in introducing new and renewable sources of energy technology (adaptation, modification, use of local materials) and the utilization of new technologies in the rural areas.

Methodologies have been developed using different approaches. Among them the promotion of self-help, or participative project implementation is the most promising although the progress has been slow, because of various reasons.

(b) Resource assessment programme

Resources assessment programmes for new and renewable sources of energy have been carried out to identify the role that such resources could play in the future and the steps that need to be taken to advance their utilization. The resources considered are: biomass, peat, micro hydropower, geothermal energy, solar, wind energy and wavepower.

The objectives of the biomass energy development policy are:

(i) Short-term/immediate objectives:

(a) To overcome the shortage of fuelwood supply; to identify wood and wood waste potential in terms of production of energy; and to demonstrate wood energy pilot plants;

(b) To provide training in wood energy technologies and management;

(c) To establish programmes of technical co-operation in the field of energy, particularly wood energy which cover the problem of supply, processing, handling, distribution, technology and demand.

(ii) Long-term objectives:

(a) To encourage people to plant wood species especially for fuelwood supply, and to help people develop fuelwood at a reasonable price;

(b) To positively contribute to the national energy plan through a diversification of the source;

(c) To speed up the national rural electrification programme and in a positive way contribute to discouraging people from leaving their areas of origin by improvement of living standards;

(d) To develop the utilization of wood into a commercial energy source.

Biomass includes 113 million ha of forest biomass, which is 59 per cent of the total land area covered by forest. In 1984 the production of wood including firewood was 165 million m³.

Agricultural waste includes crop waste (from rice, maize, cassava, sweet potatoes, peanuts, soybean, coconut, coffee); waste from plantations (mill effluents: palm oil); and animal residues. A recent assessment on the potential of agricultural waste indicated an annual amount of 29.5 million tonnes. A great number of livestock such as buffaloes, cows, etc. have been assessed to produce adequate volumes of dung for biogas production. Annual dung production amounts to about 114 million tonnes.

Owing to the importance of biomass as a source of energy, as mentioned earlier, efforts have been directed towards development of energy plantations where unproductive public land areas are available; efficient utilization of wood and biomass wastes; research and development of biomass technology (wood gasification, charcoal production, briquetting); development of liquid fuel (gasohol).

There have been intensive and continuous efforts for a regeneration of degraded land areas and for firewood supply through a reforestation programme.

Production of alcohol for gasohol has been developed. A pilot laboratory in Lampung, is in operation for the production of alcohol using sweet potatoes as feedstock. This pilot plant has a capacity of 5,000 kilolitres a year producing 95 per cent grade alcohol; together with this pilot plant an agriculture research station with a laboratory and training facilities was established with an area of 10 hectares.

There are large variations of solar energy intensity over the country. From various surveys, it is known that the average annual solar radiation in Indonesia is between 1,668 kWh/m² and 1,946 kWh/m². The average effective radiation is approximately 6 hours/day.

Wind power is still used in the traditional way such as in the traditional mode of coastal fishing and inter-island shipping. According to available data, several locations have average recorded wind speeds of 20 km/hr or more with an intensity of more than 1,500 kWh/m²/hr.

Peat is available in Sumatra and Kalimantan with resources of about 200 billion tonnes. Studies have been started to utilize peat in the near future as a substitute for firewood for domestic use and for medium- to large-size power generation.

Being an archipelago, Indonesia has the potential for ocean thermal energy conversion (OTEC). Several areas, such as Bali, have OTEC potential which can be used to generate electric power.

The country is rich in wave energy especially at the South Coast of Java and Bali and the West Coast of Sumatra. The potential will be high enough to be put to practical use. Preliminary observations show wave heights of around 2 metres with frequencies of around 9 seconds. Norway and Japan have shown their interest in co-operating in developing these potentials.

(c)   Technological assessment programme

As part of the technological assessment programme, demonstration projects have been installed in rural areas to enable people living around the units to duplicate them for further utilization of this technology. In 1979 the government installed a solar photovoltaic demonstration project of 5.5 kW peak for irrigation in the village Picon, West Java. This was followed by several other projects, among others a solar photovoltaic unit of 1.5 kW for clean water pumping in Donomulyo Village in Central Java, and a 2.5 kW unit in Maluku Province. Up to now, 53 units of PV system have been installed by the Directorate General of Electric Power and New Energy and BPPT (Agency for the Assessment and Application of Technology). Other demonstration units include wood and rice husk gasifiers, windmills and biogas plants.

Gasification units have been introduced at locations where biomass or agriculture wastes like timber mills or rice mills are available.

5. Strategies for the accelerated development of new and renewable sources of energy: the Philippine programme

Introduction

This paper discussed the Philippine Nonconventional Energy Development Programme (NEDP) with the objective of studying the issues and problems involved in the effort to increase significantly the contribution of new and renewable sources of energy to the country's energy supply.

In the Philippines, nonconventional energy resources are defined as "energy resources in which the conversion or utilization technology for large-scale application is not as well-developed and/or widely used" and that these "include the direct and indirect forms of solar, tidal, nuclear converter and breeder reactors and fusion".

In this paper, the term nonconventional energy resources is to be understood as synonymous with the term new and renewable sources of energy.

(a) Early strategies of the Philippine programme

Nonconventional technology research, development and demonstration programme

It was in 1977 when the Philippine Ministry of Energy (now Office of Energy Affairs) made a serious attempt to come up with a programme to develop noncon energy resources. This first programme was essentially a research, development and demonstration programme (R, D and D programme). Under this programme, a number of research and development studies, economic assessment and promotion activities for the whole gamut of new and emerging technologies utilizing renewable resources were implemented. The results of the various studies and projects conducted enabled the Ministry's Nonconventional Resources Devision (NCRD) to identify which of the technologies had the most potential for utilizing the available renewable energy resources in the country.

Utilization of the country's geothermal resources for power production was assessed to be already commercially viable. This led to an independent programme for the development of this resource towards increased commercial utilization. The many activities of the programme were conducted by various government agencies and corporations but were closely co-ordinated by the former Ministry of Energy.

Two other renewable energy systems, minihydro and dendro power systems, which were then perceived to be commercially viable too, were incorporated in the rural electrification programme of the now defunct Ministry of Human Settlements. The programme was 'implemented by the Ministry's National Electrification Administration (NEA) office. (The programme is still ongoing and NEA is now under the Department of Environment and Natural Resources.)

The development of the other nonconventional energy systems remained under the NCRD R, D and D programme. NCRD continued with its programme, conducting research, pilot studies, and demonstration projects on various nonconventional energy systems with the end in view of achieving technical and economic competitiveness vis-à-vis conventional energy systems. This was later coupled with a nonconventional energy promotion programme which consisted of lectures, workshops, seminars, farms, exhibits and publication and distribution of literature on various topics about nonconventional energy. The programme aimed, at increasing public awareness and interest on nonconventional energy.

Nonconventional energy commercialization programme

In 1982, five years after the NEDP was initiated by NCRD, an assessment and evaluation of the programme was conducted. This assessment and evaluation was a study that looked into the potential for commercialization of the various nonconventional energy systems under the programme and set the motion too for a new nonconventional energy development programme.

Four nonconventional energy systems were identified. These were:

(i) Biomass-fired boiler systems;

(ii) Gasifier systems for power and process heat production;

(iii) Biogas systems for power and process heat production;

(iv) Biomass-derived liquid fuel systems;

(v) Commercial-scale solar water heater systems.

(b) Present and future programme directions

At present, the emerging direction of the programme for the development of nonconventional energy utilization in the Philippines is three-pronged. First, since several nonconventional energy systems which show strong potential for substituting conventional fuels such as oil and coal need further technical development studies, NCRD will carry out a technology programme which will be a continuation of the Technology Development Programmes. There will, however, be changes in the phases composing this programme. There will be two programmes whose aims will be to conduct promotion activities to increase users of viable nonconventional energy systems; the first is the Nonconventional Energy Commercialization Programme which is to be redefined as Energy Technology Alternatives Promotions Programme or ENERTAP. The second is the Affiliated Nonconventional Centre Programme or the ANC Programme.

Basically, the previous Nonconventional Energy Commercialization Programme will be split into two phases. All activities related to R, D and D will form the Technology Programme. All non-technology development activities in the pre-commercialization phase related to encouraging the market for nonconventional energy systems will be under ENERTAP. In essence, therefore, there will be no change in terms of the policy of using essentially market forces in assigning priority to nonconventional energy systems. The main difference from the previous years is the mechanism by which the programmes will be managed.

The Technology Programme will consist of three phases; the technology Assessment phase, the technology development phase and the technology demonstration phase. Technology assessment will consist of literature research and desk study activities. Technology development will consist of pilot and field studies under controlled conditions. Technology demonstration will consist of field studies approximating actual environmental conditions to which the nonconventional energy systems will be subjected when used on a commercial scale.

ENERTAP or the Energy Technology Alternatives Promotion Programme aims to promote the utilization of nonconventional energy systems by conducting projects and activities that will address the information requirements of the two main actors in the market, the buyer and the seller. Further, it will have activities too that will institutionalize linkages required to stabilize an infantile nonconventional energy industry sector. The programme will rely primarily on the private sector, i.e. large-scale energy users both in the urban and rural areas, mostly industrial and agricultural establishments which, relatively, can afford the initial costs of putting up nonconventional energy systems.

The ANC Programme or the Affiliated Nonconventional Centre Programme is aimed at establishing a mechanism by which the energy needs in the rural areas are addressed. The energy needs in the rural areas are characterized as low-level, isolated, and of course, dispersely distributed. This is because the country is archipelagic with most of its islands ruggedly terrained. The aggregate consumption of energy in the rural areas is enormous but NCRD sees that this demand can be matched by nonconventional energy systems.

The ANC Programme therefore aims to establish long-term links with strategically located universities and colleges, so that these institutions will serve as the links of NCRD in the rural areas. Through these institutions, NCRD will identify and implement projects on nonconventional energy systems appropriate to the needs and capabilities of communities in the rural areas. NCRD, however, sees the importance of getting the participation and inputs of the affected communities in the implementation of projects. The task of the ANCs includes the identification of the projects, the putting-up of the projects and eventually the managing of the operation of the projects.

6. Use of non-traditional renewable energy sources: an important task of the national economy*
(Synopsis)

The main provisions of the Long-term USSR Energy Programme relate to the creation of a material and technical base for wide-scale use of non-traditional energy sources in the first phase of its implementation. These sources include renewable energy sources, such as the energy of solar radiation, wind energy, geothermal energy and biomass energy. These types of energy cannot compete with oil, gas, coal, or nuclear power in the near future; but, they can become an important supplementary source of energy, which should add to the reliability of power and heat supply at numerous enterprises to improve social standards in a number of regions in the USSR, and to save high-grade organic fuels (oil and gas).

---

* Mr. V.I. Dobrokhotov, USSR State Committee for Science and Technology.

There are approximately 90 million people living in regions of the USSR which are favourable from the point of view of using solar energy, including 50 million who live in the countryside.

Goals of further agricultural development require rapid adaptation of collective and state-owned farms, along with other enterprises in the agro-industrial complex, to the use of electric power. Today, there are 320 electric motors per state-owned or collective farm on average.

Renewable energy sources should increase agricultural output, and step up implementation of socio-economic reform in the countryside.

In 1985 use of non-traditional renewable energy sources (NRES) saved 0.7 million tons of organic fuel equivalent, including 0.1 million tons of fuel equivalent from the use of solar energy.

Development of new and renewable sources of energy is carried out in the following areas:

(i) Use of solar energy to generate heat for hot water supply and heating with solar collectors.

(ii) Use of solar energy for production of electric power on the basis of direct photoelectric transformation.

(iii) Geothermal energy is widely used in a number of areas, to provide consumers with heat and, to a lesser extent, electric power. At the moment 52 thermal water and vapor-water sites are being developed for over 400 consumers. Geothermal energy is used to heat houses, to provide hot water, and for central heating of apartments. Since 1967 the country's first geothermal electric power station with a design output of 11 megawatts has been operational, and work is being carried out on the construction of a geothermal electric power station with an output of 50 megawatts, which is subsequently to be raised to 200 megawatts.

(iv) Although approximately 70 per cent of the territory in the USSR is fit for year-round utilization of small wind power plants, wind energy has not been utilized at any significant level. The wind energy is harnessed via the creation of autonomous wind power plants of various capacity up to 100 kilowatts, which provide electricity, pump water, and serve other needs. One thousand 4 kW wind plants have been produced for various regions of the country in the period between 1981 and 1985. An additional eight types of wind plants having the capacity of up to 30 kW have been developed, with its serial production to begin shortly.

(v) Biomass: the potential yield of fuel from the 500 million tons of dry organic wastes, which accumulates in the country every year, can amount to nearly 150 million tons of fuel equivalent per year, provided that existing bioconversion technologies are used for this purpose. It is planned to obtain 1.7 million tons of fuel equivalent in

1990. The solution of this problem should be regarded not only from the point of view of power, but also with reference to ecology and production of high-grade organomineral fertilizers.

(vi) It is necessary to develop and build animal farms with biogas power plants. If this problem is solved today, each farm will be in a position to improve its ecological situation, not to mention having an autonomous source of power operating on locally-produced fuel which can be used in production. This is certain to upgrade the reliability of power supply to agricultural production and, at the same time, to obtain high quality fertilizers, the production of which still consumes great amounts of natural gas.

The scientific establishments of the USSR MINENERGO, the USSR MINENERGOMASH and the USSR MINELEKTROTEKHPROM have been given the task of developing new technical solutions with high techno-economic yields for harnessing the energy from tides, small rivers, waves, and others.

The USSR State Committee for Science and Technology approved the national scientific technological programme for the twelfth five-year period entitled "To elaborate and widely use progressive technology for converting solar, geothermal, wind energy and biomass to promote the use of renewable energy sources in the fuel-energy balance of the country".

Over 300 organizations attached to various ministries and departments at the national level take part in the programme.

7. Brief note on the status of Norwegian wave energy plant projects in the South Pacific*

The prototype plants

Two different types of wave energy plants have been operating since autumn 1985 at the Norwegian west coast, feeding electricity into the local grid. One of the prototypes is based on the multiresonant oscillating water column (MOWC) principle, the other on the tapered channel (TC) principle.

The development and the construction of these two prototype plants have been financed in co-operation between private industry and the Norwegian Government (Kvaerner Brug A/S, Norwave A/S and the Royal Norwegian Ministry of Oil and Energy).

Activities in the south Pacific

In co-operation with CCOP/SOPAC (Committee for Co-Ordination of Joint Prospecting for Mineral Resources in South Pacific Offshore Areas), an intergovernmental

organization comprising 12 member countries in the South Pacific region, a wave energy programme was started in 1984.

Interested member countries were visited by a wave energy utilization mission which, among others, recommended favourable sites as regards wave energy potential. Wave measurement equipment including wave rider bouys and satellite communication equipment were purchased, and stations set up at the south-west coast of Tongatapu Island (Tonga) and at Rarotonga Island (Cook Islands). Prefeasibility studies of favourable construction sites have been carried out including cost estimates for construction of a prototype plant in the region. Rough estimate is about $US1.9 million excluding cost of the power-generating equipment.

The Norwegian support to the CCOP/SOPAC programme is granted by the Ministry for Development Co-operation (DUH). The co-operation from the Norwegian side is co-ordinated by NECOR Foundation, which is the appointed national counterpart agency to CCOP/SOPAC.

Evaluations of the prototype plants

Extensive testing of the two prototype plants has been carried out during a period of about two years. A committee has been appointed by the Norwegian Ministry of Oil and Energy (OED) which is expected to finalize its report by the end of 1987. The task of this Committee is, in short, to sum up the experience obtained, to compare ideal construction costs of the two plants and to compare energy conversion efficiencies and kWh cost.

When, hopefully, the first prototype plant is put into operation on a South Pacific island sometime in the near future, it has to make sure that the country and community it serves will have full benefit of it. A next step in the evaluation process, therefore, is to determine the usefulness, i.e. the efficiency and operational reliability of the plant, for use in a developing country. To perform this task an independent highly qualified body will be appointed by the NECOR Foundation. The work will to a large extent be based on the evaluations carried out by OED, and should be completed in early 1988.

Further Norwegian Government support to the CCOP/SOPAC wave energy programme will depend on the result of this independent verification.

More information of the projects
may be obtained from:

NECOR Foundation
Veritasveien 1, P.O. Box 350
1322 HOEVIK
Norway

Tel.: (02) 47 99 00
Telex: 76 192 verit n

---

* Presented by the Norwegian Engineering Council on Oceanic Resources (NECOR) Foundation.

Annex

LIST OF DOCUMENTS PRESENTED TO THE COMMITTEE ON NATURAL RESOURCES AT ITS FOURTEENTH SESSION

	Title	*Source*
1. Secretariat documents		
E/ESCAP/NR.14/L.1	Provisional agenda	Secretariat
E/ESCAP/NR.14/L.2	Annotated provisional agenda	Secretariat
E/ESCAP/NR.14/1	Issues related to the regional economy pattern	Secretariat
E/ESCAP/NR.14/2	Effects of price and non-price policies on energy demand management	Secretariat
E/ESCAP/NR.14/3	The use of coal in households and small-scale industries	Secretariat
E/ESCAP/NR.14/4	Report of the Meeting of Senior Experts Preparatory to the Fourteenth Session of the Committee on Natural Resources	Secretariat
E/ESCAP/NR.14/5	Updated assessment of the contribution of new and renewable sources of energy to regional energy supply	Secretariat
E/ESCAP/NR.14/6	Co-operative research, development and demonstration achievements and future plans regarding new and renewable sources of energy	Secretariat
E/ESCAP/NR.14/7 and Corr.1	Potential for Asian trans-country power exchange and development	Secretariat
E/ESCAP/NR.14/8	Human resources development issues	Secretariat
E/ESCAP/NR.14/9	The regional energy scene	Secretariat
E/ESCAP/NR.14/10	Prospects for production and utilization of coal, natural gas and electricity	Secretariat
E/ESCAP/NR.14/11	Draft medium-term plan, 1990-1995; and programme changes, 1988-1989: energy resources	Secretariat
E/ESCAP/NR.14/12	Activities of ESCAP in regard to natural resources: water resources	Secretariat
E/ESCAP/NR.14/13	Rain-water harvesting techniques and prospects for their application in developing island countries	Secretariat
E/ESCAP/NR.14/14	Activities of ESCAP in regard to natural resources: mineral resources	Secretariat
E/ESCAP/NR.14/15	Activities of ESCAP in regard to natural resources: cartography and remote sensing	Secretariat
E/ESCAP/NR.14/16	Water tariffs as a policy instrument to provide better management of water resources	Secretariat
E/ESCAP/NR.14/17	Activities of ESCAP in regard to natural resources: marine resources	Secretariat
E/ESCAP/NR.14/18	Activities of ESCAP in regard to natural resources: energy resources	Secretariat
E/ESCAP/NR.14/19	Programme changes, 1988-1989; and proposed draft medium-term plan, 1990-1995: marine resources	Secretariat

E/ESCAP/NR.14/20	Regional mineral resources development programme	Secretariat
E/ESCAP/NR.14/21	Safety evaluation of existing dams	Secretariat
E/ESCAP/NR.14/22 and Add.1	Draft medium-term plan, 1990-1995; and programme changes, 1988-1989: natural resources	Secretariat
E/ESCAP/NR.14/23	Consideration of the agenda and arrangements for subsequent sessions of the Committee	Secretariat
—	Report of the tripartite review conference of the regional energy development programme	REDP

2. Papers prepared by other organizations

—	Summary of technical co-operation activities in energy	Department of Technical Co-operation for Development
—	Technical co-operation programme in energy resources development	
—	Present emphasis of ECE energy activities and available documents	ECE
—	Energy research: directions and issues for developing countries	ERG/IDRC
—	The energy research group and the Asia-Pacific region	ERG/IDRG
—	FAO activities concerning development of renewable sources of rural energy in the Asia-Pacific region	FAO
—	Current activities of the International Energy Agency	IEA
—	Energy manpower planning, training and related socio-economic issues: an ILO perspective	ILO
—	General information bulletin	SIAP
—	SPEC energy activities	SPEC
	Regional Energy Meeting, Honiara, Solomon Islands	
—	Information note by the World Energy Conference	WEC
—	The World Bank Annual Report 1987	World Bank

3. Country papers

—	Country paper	Bangladesh
—	Prospects of the production and use of coal and natural gas and power generation in China	China
	The production, consumption and policy of energy resources in China	
—	Efforts to improve the utilization of new and renewable sources of energy	Indonesia
	Human resources development in Indonesia	
	The energy development in Indonesia	
	Future trends of electric power in Indonesia	
	Prospects of coal mining development in Indonesia	
	Hydropower in Indonesia	

−	Brief note on the status on Norwegion ocean mining efforts	Norway
	Brief note on the status of Norwegian wave energy plant projects in the South Pacific	
−	The Philippine energy scenario	Philippines
	The Philippine case	
	Strategies for the accelerated development of new and renewable sources of energy: the Philippine program	
−	Energy issue: prospects for production and utilization of natural gas in Thailand	Thailand
−	Use of non-traditional renewable energy sources: an important task of the national economy	USSR